SCIENCE WITH COMPUTERS

SCIENCE WITH COMPUTERS

RICHARD C. ADAMS

AN EXPERIMENTAL
SCIENCE SERIES BOOK

A GROLIER COMPANY

FRANKLIN WATTS
NEW YORK/LONDON
TORONTO/SYDNEY/1987

Diagrams by Vantage Art

All photographs courtesy of the author

Library of Congress Cataloging-in-Publication Data

Adams, Richard Craig.
Science with computers.

(An Experimental science series book)
Bibliography: p.
Includes index.
Summary: Discusses how to simplify science projects by using computers to organize data, do calculations, or suggest new avenues of study.
1. Science—Data Processing—Juvenile literature.
[1. Science—Data Processing] I. Title. II. Series.
Q183.9.A33 1987 502.8′5 86-26725
ISBN 0-531-10324-2

Printed in the United States of America
6 5 4 3 2 1

CONTENTS

SCIENCE WITH COMPUTERS

INTRODUCTION

You like computers and you like science.

You wouldn't have picked up this book unless you fit that description just a little bit. It would be nice if you could combine your interests.

Of course, it may be that you don't really think you like science and were just intrigued by the computers part. Don't be too sure about your dislikes. Perhaps you don't know what science really is. All kids are natural-born scientists. They look carefully at their world, make guesses about how it operates, and then try out their guesses to see if they work. In everything they do, they are delighted in the newness of the things they experience—just like scientists.

Sometimes, overcrowded schools and harried teachers can drive this curiosity out of you by making a science class seem to be only learning "scientific laws," the history of famous scientists, and how to fill out lab report forms in neat handwriting in ink. Occasionally, it seems that students can be discouraged from asking "foolish questions."

But no question you ask is foolish. It's like telling somebody who is afraid of spiders to stop having "unreasonable fears." Any fear you have is quite reasonable TO YOU. Any question you have is valid because you want to know the answer. Sometimes the answer is not what you expected it to be, but that is when you start to learn.

No teacher assigns or allows students to do science projects to make his or her job easier. Projects are a real headache to get started, administer, nurse along during the difficult times, and prepare for presentation. Why do so many teachers enjoy them, and why do so many kids like them after they get finished? Because the kids are not just *learning* about science. They are DOING science.

Computers can take some of the pain out of doing a very good job on a science project. They can make your data easier to handle, the mathematics a whiz, the results more valid and easier to interpret; they can suggest new avenues of study and make your final report clearer and easier to do.

YOU still have to come up with the project, although I will give a few suggestions. YOU still have to do the work, but the computer can save you from the repetitious arithmetic.

Science is a human activity. Computers just allow you to be more human while doing it.

A NOTE ON THE PROGRAMS IN THIS BOOK

Most of the programs in this book were written to run on the Apple II series of computers (Apple II+, IIe, or IIc). If you are in a public school, there is an 87% chance that you will have access to an Apple and be able to type in these programs as they are.

The computer language BASIC is offered on all microcomputers, and many lines of the programs here will be the same for all computers. The real differences will come in graphics and special things. I have tried to keep the programs simple, except when there was no way to do that without making it terribly dull. I will be giving some examples for the Commodore 64, Atari, and Radio Shack computers as I go along.

Although this is not a book on programming, I have tried to make the programs as easy to understand as I could. Many students pick up some BASIC programming on their own and may not learn the best way of doing things. BASIC is a forgiving enough computer language to let you do things in a sloppy manner and still have a program that works.

Many beginning programmers will try to use variable names in their programs that are as short as possible to avoid typing. However, that doesn't make it easy to tell what you did in a program even a couple of weeks later. For example, in figuring out an average, a beginning programmer might write:

```
30 A = X1 / N
```

My programs would do that calculation in a line like this:

```
30 AVERAGE = SUM / NUMBER
```

This takes more typing, but it is much easier to understand what is happening. There are a few cautions about this technique, however. Computers do not always look at the entire word. The Apple, Radio Shack, and Commodore, for example, look only at the first two letters. Thus, AVERAGE and AVOIDANCE would be seen as the same name. The Atari sees the first 128 characters and so would not have this problem.

Another thing is to see if your computer has "reserved" words. If you tried to type in this line on an Apple:

```
30 AVERAGE = TOTAL/RECORDS
```

the LISTing of the line would appear as:

```
30 AVERAGE = TOTAL / RECORDS
```

and RUNning the program would produce a beep and SYNTAX ERROR. The computer reserves the words TO and OR for its own commands, and you can't write names that use them. The usual solution is to use a different word or one that is close, such as TTAL and RECRDS.

Lastly, there are several ways to LIST programs. If you list a program on the Apple, it will have a format something like that of this program:

```
10 FOR ENTRY = 1 TO NUMBER

20 PRINT "AND A "; : PRINT ENTRY : PRINT
   : PRINT ENTRY + 100 : X = ENTRY + 20

30 IF ENTRY = 10 THEN PRINT "THAT WAS 10"
   : PRINT "HOORAY"

40 NEXT ENTRY
```

In order to save space, the same listing in this book would eliminate the indentations that occur on the second lines of lines 20 and 30, and would not have any extra lines between entries. This makes it harder to read, but we were able to get more programs into the book that way.

Whatever computer you use in your science project, it can help you in many ways, even if you never type in one of the programs in this book. You can always ask for help from a teacher or one of the other students to teach you how to run the computer and commercial software or programs others have done.

One final note. If you don't trust your ability to type in all these programs, you can get them already on Apple and Atari disks. The Apple disk will also contain a number of other programs, including the one for the three-dimensional bar graph shown in Chapter 2.

The Apple disk is available for $15.00 + $1.25 shipping, and the Atari disk for $20.00 + $1.25 shipping, from Adams' Soft Enterprises, P.O. Box 386, Springfield, OR 97477.

1 SCIENTISTS AND COMPUTERS

It was not a backyard tinkerer who invented the first computer. Credit for that must go to Blaise Pascal, a French mathematician; Charles Babbage, who designed a "difference engine"; and Lady Ada Lovelace, the person who first designed programs to use with Babbage's machine, which would have taken up an acre of space had it ever been built.

From the beginning, scientists and mathematicians were the ones who designed and built computers. They had the ideas, they had the skills, they had the *need* for the power that computers could give them.

Of course, it WAS a pair of tinkerers in a garage near San Francisco who started the microcomputer industry when they showed their "Apple" to the Homebrew Computer Club. Still, as long as microcomputers have been available, scientists have been designing programs, devices, and peripherals to make these machines do what the scientists need to further their work.

What do scientists DO with computers?

Scientists first used computers as a mathematical

tool, the way they had adopted the slide rule, adding machine, and later the calculator. This was just aiding them in what they normally did. The only difference was that now they got done with the tedious arithmetic sooner and could get onto new phases of a project.

Later, by lowering the effective cost of calculating compared with experimenting, the computer helped make a shift toward calculation and simulation in many fields where once only experimentation and comparatively direct measurement were practical. A **simulation** is a program that makes the computer give numerical or graphical results the same way the actual process would. These allow predictions of how the real-life process would function with changed conditions. Even though good simulations with complex mathematics are now easier, one of the primary uses of computers is as **smart tools.** You'll understand what a smart tool is by reading the following example.

In a certain experiment, a scientist has to record temperatures. The scientist reads a thermometer at specific times and records the data in a notebook. Later on, the scientist may make a graph of these numbers as they varied over time or perform calculations with the temperature changes.

A **thermocouple** is a device which indicates temperatures by changing how electricity flows through its wires. A computer is a smart tool when you hook up a thermocouple to it and use it instead of a thermometer in the lab. Not only will the computer record the changing temperatures at various times for you, it can also be doing calculations and even graphing the numbers at the same time.

The computer as a smart tool measures the data very accurately and also processes it so that the scientist can use it more easily.

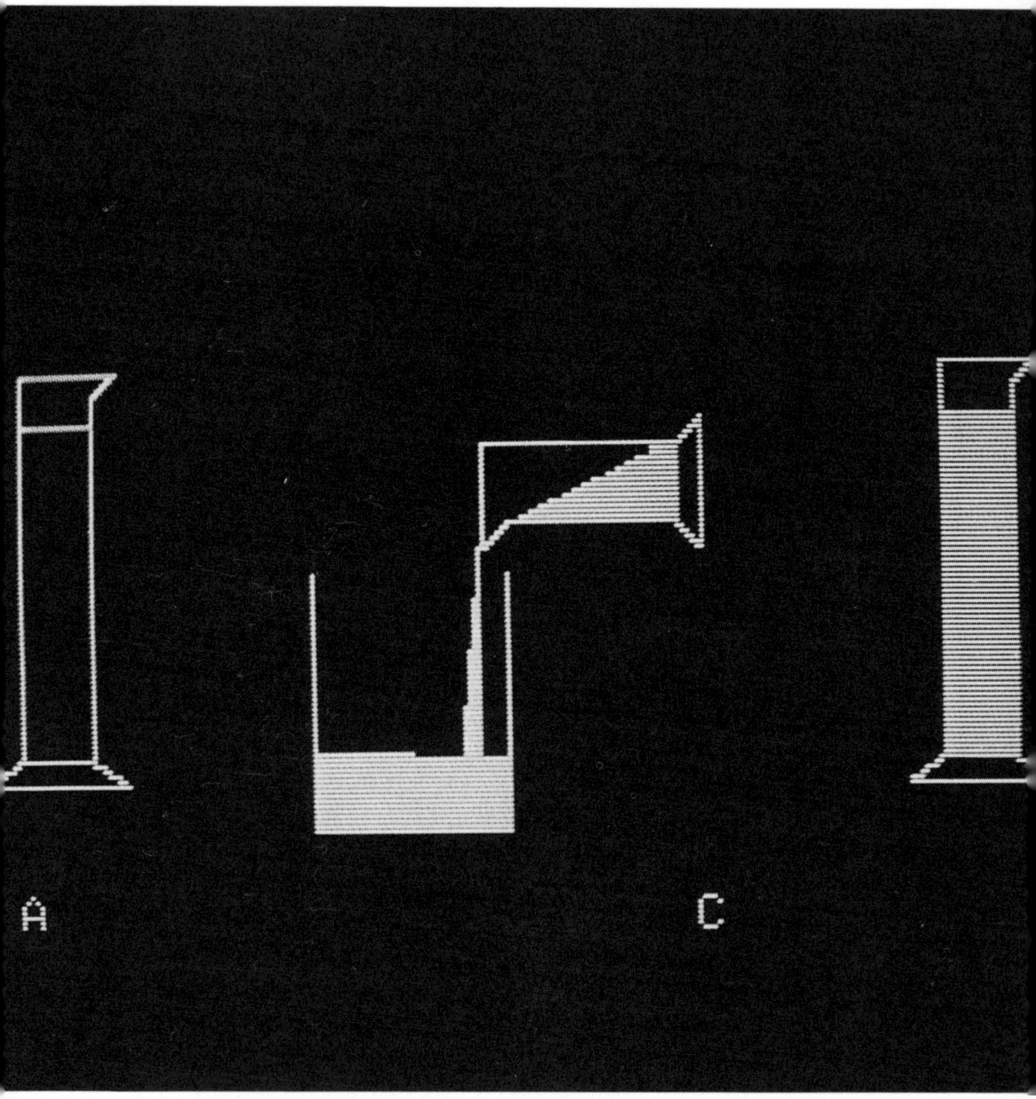

Computer simulation of a chemistry experiment. The software is Chemical Kinetics *from Ken Hartman.*

This processing of information is a major use for computers. There are some studies, especially those dealing with large numbers of examples, such as collections of molecules, atoms, bacteria, or people, that come up with numbers just too big to handle easily. **Statistics** was developed as a method of dealing with large numbers of cases and getting useful results. By using its techniques, you can't really tell what any one atom, molecule, or person will do, but you have a fairly good idea of what large numbers of them will do. As better techniques were developed, the problem then became the tremendous number of calculations that had to be done. Some of these were very complex and doing even a few of them was difficult. The computer allowed many complex calculations to be done quickly, letting scientists spend their energies on interpreting the results and asking new questions.

Finally, the computer has aided the scientist in communicating his or her findings to the rest of the scientific community and to the public at large. Take the graphics capabilities of the computer. A computer can quickly graph data. And it can show animated pictures of shuttle missions, Jupiter and Saturn flybys, models of atoms in motion, and things that happen too fast or too far away to see. This ability has led to a better understanding of scientific ideas by many more people than previously.

Word processing, and quickly made charts, graphs, and pictures have aided publication of findings so that results are known much more quickly. Communication between scientists about experiments allows many more minds to work on the same problem: to verify what has been done for accuracy, to see different ways to approach the problem than have been tried so far, and to puzzle over the things that went wrong and figure out what they truly mean.

The computer has given the scientist the gift of time and the aid of a skilled assistant. The computer has not increased the skill of the scientist, but instead has allowed him or her to spend more time in the creative aspects of science—asking the questions and thinking of possible answers.

A truly fine project, a hobby, or even a career may start with you, a computer, and one good idea.

2
NUMBER CRUNCHING

Early humans counted on their fingers. Later peoples used such devices as the abacus, with beads strung on wires. The Romans used the same idea but made a board with grooves in which they moved smooth stones. The Latin word for "stones" is *calcis*. From the name of the integral parts of this counting board comes our word *calculate*, and we call our counting devices *calculators*.

Or *computers*.

How is a computer different from a calculator? The main difference is that a computer can be told to do a *sequence of actions*, while a calculator can do only one thing at a time.

A computer can work only with numbers to do all the amazing things it does. Even when the computer looks as if it is working with words, it is still doing numbers. Here, try this Apple program:

```
10 FOR NUMBER = 65 TO 91
20 PRINT CHR$(NUMBER);CHR$(32);
30 NEXT NUMBER
40 PRINT CHR$(72) + CHR$(73) + CHR$(33)
50 END
```

Nowhere in that program did I tell the computer to print anything but numbers, yet it prints letters. In line 40, I even had the computer *adding* these things together. The computer can use the concepts of addition, subtraction, and even "greater than" and "less than" with letters, since it is really working with numbers. The computer has a "look-up list" of letters that it can substitute for numbers, called the American Standard Code for Information Interchange, usually abbreviated ASCII.

Still, the computer really does do number stuff quite well. The first computers built in this country were used to figure out tables of numbers to be used by artillery officers in aiming their big cannons.

THE COMPUTER AS CALCULATOR

You can use the computer as a calculator or even write programs to do slightly more complicated things. For example, to make most computers with BASIC do a calculation, such as 6 + 4, all you need to do is type PRINT 6 + 4 and press the RETURN or ENTER key. On the next line will be 10. Sure, it's not quite as easy as a calculator, but then, if you don't have a calculator handy and the computer IS there, it's easier than counting on your fingers, especially for long division.

Suppose you had to find out the area of a circle. You could use the formula you have learned: $A = 3.14 r^2$, where 3.14 is pi and r^2 is the radius squared. But what if you had a whole bunch of areas of circles to figure out? Then the computer would make it much easier than a calculator. Try this program:

```
10  REM  AREA OF A CIRCLE
20  PI = 3.14159
30  PRINT "THIS PROGRAM WILL FIND THE AREA OF
```

```
A" : PRINT "CIRCLE" : PRINT
40  PRINT "WHAT IS THE RADIUS"; : INPUT RADIUS
50  AREA = PI * RADIUS ^ 2
60  PRINT : PRINT "THE AREA OF THAT CIRCLE IS
";AREA
70  PRINT : PRINT "ANOTHER? (Y/N)"; : GET
ANSWER$ : PRINT ANSWER$ : IF ANSWER$ <  > "Y"
AND ANSWER$ <  > "N" THEN 70
80  PRINT : PRINT
90  IF ANSWER$ = "Y" THEN 40
100  END
```

Line 50 is where I do the calculation. The ^ is how most computers show "power," as in the second power or "squared." Line 60 tells the answer. One of the most important parts of this program is line 90. If you want to do another circle, that line tells the computer to go back to the beginning and do it again. This is the real power of the computer: fast calculation done over and over.

HOW MANY PLACES?

Something you may notice is that the computer doesn't carry out numbers more than nine decimal places (at least on the Apple). Some computers do more places, some less. What if you want it to do more digits than that? You CAN make your computer give you as many places, say, in a division, as you want, just by imitating the way YOU do long division.

In the program below, the numbers are divided and the first nine numbers put down. The remainder is divided and the next nine numbers written down. Then its remainder is again divided and so on. I've made the program break up the numbers into groups of three, to make it easier to see how many places there are. Using spaces instead of commas to break up number groups into three is the way things are done in international scientific journals, since some countries use commas for decimals and periods for number grouping.

Try writing the program below to get some truly extraordinary numbers that will impress anyone.

```
10  REM  MULTIPLE PRECISION DIVIDER
20  DEF FN CHECKSIZE(X) =  LOG (X) / LOG (10)
30  D$ =  CHR$ (4) : USUALDIGITS = 9 : REM
"Largest size number the machine can do without
going into scientific notation. See REMark on
line 260
50  GOSUB 100 : REM  GET THE NUMBER FIRST
60  GOSUB 200 : REM  NOW WORK ON IT
70  PRINT "Do you want to do another?"; : GET
ANSWER$ : PRINT ANSWER$ : IF ANSWER$ <  > "Y"
AND ANSWER$ <  > "N" THEN 70
80  IF ANSWER$ = "Y" THEN 50
90  END
100  REM  GET THE NUMBERS FIRST
105  HOME : PRINT "This will give you a
division carried" : PRINT "out to as many
places as you want."
110  PRINT : PRINT : PRINT "What number do you
wish to divide" : INPUT DIVIDND : DIVIDND$ =
STR$ (DIVIDND) : IF LEN (DIVIDND$) >
USUALDIGITS THEN GOSUB 190 : GOTO 110
120  PRINT : PRINT "And what do you want to
divide " : PRINT DIVIDND;" by"; : INPUT DVISR :
DVISR$ = STR$ (DVISR) : IF LEN (DVISR$) >
USUALDIGITS THEN GOSUB 190 : GOTO 120
130  QUOTIENT = DIVIDND / DVISR : IF QUOTIENT >
1E + 09 OR QUOTIENT < 1E - 09 THEN PRINT :
PRINT "I can't do two numbers that far apart."
: PRINT "Try again."; : GET KEYPRESS$ : PRINT :
GOTO 110
140  PRINT : PRINT "O.K., now how many places
do you need" : PRINT "to have the answer
```

```
carried out"; : INPUT NUMBER
150  PRINT : PRINT "Lastly, do you want the
answer to go to" : PRINT : PRINT "(S)creen" :
PRINT "(P)rinter"; : GET PO$ : PRINT PO$ : IF
PO$ <  > "S" AND PO$ <  > "P" THEN 150
160  IF PO$ = "P" THEN PRINT D$;"PR#1"
180  PRINT DIVIDND;" divided by ";DVISR;"," :
PRINT "to at least ";NUMBER;" places, is:" :
PRINT : RETURN
190  PRINT "I can only handle nine-digit
numbers," : PRINT "excluding the decimal point.
Try again." : GET KEYPRESS$ : PRINT : RETURN
200  REM  DO THE CALCULATION
210  NUMBER = NUMBER / USUALDIGITS : REM  Print
out the parts of the number in groups of nine
digits or whatever
220  QUOTIENT =  INT (DIVIDND / DVISR)
230  LET COUNT = USUALDIGITS
240  IF QUOTIENT = 0 THEN  PRINT "0.";
250  IF QUOTIENT <  > 0 THEN  PRINT
QUOTIENT;"."; : REM  "240 & 250 print part of
the number to the left of the decimal
259  REM  LONG DIVISION STEPS
260  LET DIVIDND = (DIVIDND - (QUOTIENT *
DVISR)) * 1.0E + 09 : REM  "You can't use the
variable USUALDIGIT here.  You have to write
this one in for your machine.
270  QUOTIENT = INT (DIVIDND / DVISR) : REM
"Divide remainder and so on...
280  IF QUOTIENT = 0 THEN  PRINT : PRINT "All
the remaining digits are zero" : GOTO 350
290  RMAINDER = INT (FN CHECKSIZE(QUOTIENT)) +
1
299  REM  300 prints out leading zeros for
numbers less than one.  The end of 300 and 320
formats the digits into groups of three, as in
international scientific usage
300  IF RMAINDER <  = (USUALDIGITS - 1) THEN
FOR ZERO = 1 TO (USUALDIGITS - RMAINDER) :
PRINT "0"; : NEXT ZERO : QU$ = STR$ (QUOTIENT)
```

```
: FOR PART = 1 TO 7 STEP 3 : PRINT MID$
(QU$,PART,3);" "; : NEXT PART
310  IF RMAINDER < = (USUALDIGITS - 1) THEN
330
320  QUOTIENT$ =  STR$ (QUOTIENT) : FOR PART =
1 TO 7 STEP 3 : PRINT MID$ (QUOTIENT$,PART,3);"
"; : NEXT PART : REM  "This formats answer into
groups of three, as in standard international
scientific usage
330  COUNT = COUNT - 1 : IF COUNT = 0 THEN
COUNT = USUALDIGITS
340  NUMBER = NUMBER - 1 : IF NUMBER > 0 THEN
260
350  PRINT : PRINT : PRINT : IF PO$ = "P" THEN
PRINT D$;"PR#0" : REM  FINISHED WITH PRINTOUT
360  RETURN
```

ALGORITHMS AND PROGRAMS

In writing a program for the computer, you first have to find the proper **algorithm**. An algorithm is the way you figure something out. The algorithm for the program to find the area of a circle consisted of three steps: (1) Find the radius. (2) Square the radius and multiply it by pi. (3) Tell the answer.

The algorithm for the multiple precision divide program was the method you use for long division. In writing a program to figure out problems, you have to be able to figure out the *method* first. The computer is stupid but fast. It cannot do anything until you tell it how, step-by-step, to do something. This set of instructions is called a **program**.

Now that you have a program to do many decimal places in a division, you might play with it and explore some numbers. For a lot of divisions, there comes a place where the answer starts to repeat, or it might actually come out even.

I tried dividing 345 by 124. It seems to start repeating after the twenty-second digit. I also tried dividing 345 by 128 and told the computer to take it out to 200 places. The computer "said" that after the ninth place past the decimal, all the rest of the digits were zero. Try it and see if you can come up with any pattern. This could be the heart of a math project in number theory.

SIGNIFICANT FIGURES

When using data from an actual experiment, watch out that you don't carry out calculations to *too many* places.

For example, suppose you are finding the density of some rock samples as an aid to identifying them. Since density is mass divided by volume, you measure the mass on a good balance and find one sample to be 144.56 grams (g). You then add water to a graduated cylinder until the level is at 100 milliliters (ml), plunk in your rock, and see that the final volume is 135 ml. You figure then that the volume has to be 35 ml.

When you do the division, the computer or calculator says the density is 4.1302857 g/ml. You had five digits from your balance, because it was quite accurate. The volume measurement, however, is only accurate to *two* figures. Any calculation you do with these two numbers can't be any more accurate than your *least accurate* measurement. This lower number of digits is called the number of **significant figures**. Thus, the *realistic* answer is not 4.1302857, but 4.1. There are rules for figuring out some special cases, which you can find in any chemistry or physics textbook and in some general science books. Be careful of your calculations involving numbers from real experiments.

An example of using the quickness of the computer to do things you could do yourself is in the quadratic equation. If you have come across this in your math class, it's for solving problems of the form like this:

$$\mathbf{Ax^2 + Bx + C = 0} \tag{1}$$

using the formula:

$$\mathbf{x = \frac{-B \pm \sqrt{B^2 - 4AC}}{2A}} \tag{2}$$

This is called the *quadratic formula*. Solving that equation is a chore, but you do it in math class to show that you can do it. In physics, however, being able to solve that sort of problem *can* be interesting when it applies to some physical event.

If you drop an object, it will *accelerate*. At the end of the first second, it will be going 32 feet per second (ft/sec), which is about 22 miles per hour (mph) or 9.8 meters per second (m/sec). At the end of the second second of falling, it will be traveling at 64 ft/sec (42 mph or 19.6 m/sec) and at the end of the third, 96 ft/sec (63 mph or 29.4 m/sec).

The formula to tell how fast an object falls when it is accelerated is velocity = acceleration × time. The acceleration for earth's gravity is 32 feet per second per second (9.8 meters per second per second). How far will an object fall if it falls 5 seconds?

To find that out takes another formula:

$$\mathbf{d = V_1t + 0.5at^2,} \tag{3}$$

where **d** is the distance fallen, $\mathbf{V_1}$ is the starting velocity, **a** is the acceleration, and **t** is the time of falling. If we throw an object down with a starting velocity of 10 ft/sec, at the end of 5 sec it should have fallen 450 ft (13.7 m).

What if I want to ask the question "How long will it take the object to fall a mile (5,280 ft, or 1.6 km)?" Rearranging the equation above, I get:

$$\mathbf{0.5at^2 + Vt - d = 0} \qquad (4)$$

But look at this: $\mathbf{Ax^2 + Bx - C = 0}$. This is equation 1 from above. I think you can see that equations 1 and 4 are really the same equations. The falling-rock equation is just in terms of **t** instead of **x**. The coefficients for our problem are $\mathbf{A = 0.5 \times 32}$, $\mathbf{B = 10}$, and $\mathbf{C = -5280}$. This could be solved using the quadratic formula above, but grinding that sort of thing out takes a lot of time and can lead to mistakes, even with a calculator. On the other hand, figuring out problems like this is very useful.

If you are in a plane dropping something, such as a load of fire-retardant on a forest fire, a life raft to someone in the water, or even a bomb, you have to know how long it takes for something to fall. If you are flying at 5,280 ft (1 mi, 1.6 km), you need to know how many seconds away from the target you are to throw out the object so that it will hit the target just as you fly over it. Since we're just dropping the object, instead of throwing it, **B** is zero, and **A** is 0.5 times 32 or 16, so our calculation becomes:

$$\mathbf{x = -\frac{0 + \sqrt{0^2 - (4)(16)(-5280)}}{(2)(16)}}$$

$$\mathbf{= \frac{-0 + \sqrt{0 + 337920}}{32}}$$

$$\mathbf{= \frac{581.31}{32} = 18.2\ sec}$$

In our case, you would have to be 18.2 sec from the target when you dropped your object.

A computer program makes this sort of problem easy. Here's one below:

```
10  REM  QUADRATIC EQUATION SOLVER
20  GOSUB 100 : REM  INTRODUCTION
30  GOSUB 200 : REM  GET THE INFORMATION
40  GOSUB 300 : REM  DO CALCULATIONS AND PRINT
RESULTS
50  IF ANSWER$ = "Y" OR ANSWER$ = "y" THEN 30
60  END
100  REM  INTRODUCTION
110  HOME : PRINT "Quadratic equations look
like this:"
120  PRINT : PRINT TAB( 5);"Ax^2 + Bx + C = 0"
: PRINT : PRINT "where 'x^2'     means 'x
squared'."
130  PRINT : PRINT : PRINT "You need to tell me
A, B, and C"
140  RETURN
200  REM  GET THE INFORMATION
210  PRINT : PRINT "What are your values for A,
B, C"
220  INPUT A,B,C
230  RETURN
300  REM  DO CALCULATIONS AND PRINT RESULTS
310  RADICAL = B ^ 2 - (4 * A * C)
320  IF RADICAL < 1 THEN  PRINT "The solutions
are  imaginary'" : GOTO 360
330  X1 = (- B +  SQR (RADICAL)) / (2 * A)
340  X2 = (- B -  SQR (RADICAL)) / (2 * A)
350  PRINT : PRINT "X = ";X1;" or  X = ";X2
360  PRINT : PRINT "another (Y/N)"; : GET
ANSWER$ : PRINT ANSWER$
370  RETURN
```

The formula works but has to ignore wind resistance for simplicity's sake. A computer could handle the extra details easily, however.

PATTERNS

Humans are spoken of in biology as unspecialized animals. We can't do any one physical skill better than any animal, but we can do a pretty good job of many things.

One thing humans have that no other animal has is very sharp color vision, coupled with manipulative organs (hands) capable of extremely precise movement.

Along with this, humans have a specialty of *finding patterns*. Our brain, if it can find a pattern, can predict what will happen next, given a certain set of circumstances. Given this power, we can construct ways to get around our physical shortcomings and do better than the specialist animals.

A baby learns to talk by finding the pattern of sound and the actions that go with them in its parents. If you were placed in a foreign country, not knowing the language, you would use the same techniques to learn how to communicate after a while. If you see a section of a brick wall, you can deduce the pattern and could keep on laying bricks the same way if you had to.

Actually, all animals are pattern finders, but usually with much less breadth of expertise. Dogs use smell, porpoises and bats sound, spiders the sense of touch, bees see patterns in ultraviolet light on flowers, and so on.

Humans can find patterns in many of these fields but may have trouble in some. One of the worst is numbers. Numbers are a rather late invention of humans. If we can turn numbers into something that stimulates one of our other senses, it becomes far easier for our brain to use the old skills it has developed over several million years. The tones you hear in the earpiece of a tone phone are pleasant but also will warn you if you make a mistake. You may not consciously sit down to memorize the proper tones for each number, but calling a good friend repeatedly will train your ear to know when the number is the "right tune" and when you have made a mistake.

In speaking, we all communicate with changes in pitch and loudness in our voices to give our listener more information that is hard to express just with words alone. In writing, we can add similar information by using boldface letters, all capitals, underlining, or italics.

For Chinese speakers, there are several ways to say each word, using tones, and the meaning will be different. English speakers use tone patterns in the *sentences*, rather than the words. Think how you end a sentence that's a question.

Color is a great aid to visual pattern-finding in humans. Most mammals can't see color; among the mammals only humans and the primates, such as chimpanzees and gorillas, have this sense. Birds and insects seem to be able to sense colors, however.

Suppose I wrote a whole bunch of numbers on a blackboard in front of you and told you to pick out all the ones that ended in 5. You would have to look at each number closely and then tell me where they were. On the other hand, what if I made all the 5's red, the 3's blue, the 7's orange, and so on? You could then instantly have a picture in your mind of where the 5's were. If they were all in the center of the blackboard, you would know it immediately. If they were only on the edges, you could see that too.

The human eye and brain are expert at seeing visual patterns. The computer can be a way in science of turning numerical data into something in which it is much easier to detect patterns.

GRAPHING

Suppose I give you a series of numbers that look like this:

78 123 89 66 93 167 178 56 95 108 88 73

What are these numbers? I spent an hour one Saturday writing down the months cars were licensed, as indicated by the little stickers on their license plates. These numbers are how many cars I found in each month group. Can you see any pattern in these numbers? Take a look at Figure 1.

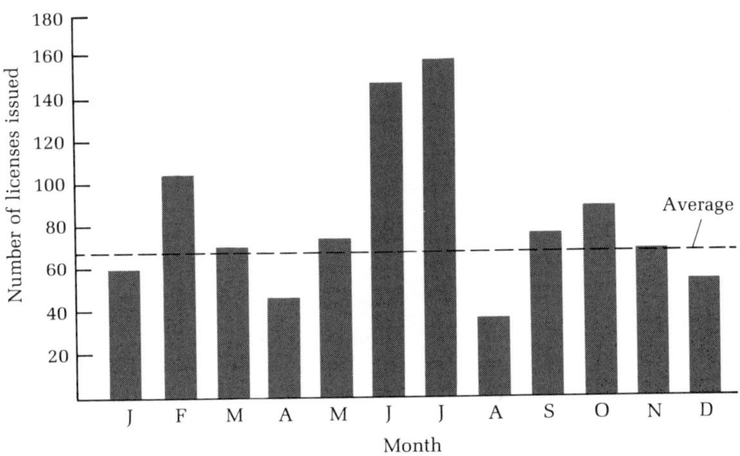

Figure 1. Graph of number of automobile licenses issued per month

Now you can see some interesting things. The letters down at the bottom stand for the months, J for January, F for February, and so on. What time of the year seems to be the most popular for buying a car? It looks like June and July. When do fewer people buy cars? From the graph, it seems that April, August, and December are not too popular. Why is that? Perhaps you can come up with some likely reasons.

December, you might say, is when people are spending their money on Christmas gifts, not cars. September is when the new models come out so maybe people are waiting in August. April is when income tax returns have to be in so it could be that people have other things on their minds. Why are June and July so popular? A lot of people take long trips during their vacations during the summer. They may feel that their old car isn't reliable enough anymore and so think they should get something new. Or maybe buying a car in the sunshine is more attractive than when the streets are frozen or slushy with snow.

Automobile manufacturers would like to find the answers to these questions and would do research like this. I used it as an example to show that numerical data don't really help you much if you can't see that a pattern even exists.

A LOW-RESOLUTION GRAPHING PROGRAM

Let's write a program to graph any numbers you put in. I'll make this first one a bar graph, using the low-resolution capabilities of the Apple.

```
10  REM  LO-RES DATA GRAPHER
20  XHIGH = 38 : YHIGH = 38 : XLOW = 1 : YLOW =
1 : ALLOWED = 38 : WHITE = 15
```

```
30  MAX = 0 : SUM = 0 : OVER = XLOW
60  GOSUB 100 : REM  GET THE DATA
70  GOSUB 200 : REM  DRAW THE GRAPH
80  END
100  REM  GET THE DATA
110  TEXT : HOME : PRINT "THIS PROGRAM CAN
GRAPH UP TO ";ALLOWED;" NUMBERS" : PRINT :
PRINT "HOW MANY DO YOU WANT TO DO"; :
INPUTNUMBER : IF NUMBER < 1 OR NUMBER > ALLOWED
THEN 110
120  DIM AMOUNT(NUMBER + 1) : PRINT
130  FOR ENTRY = 1 TO NUMBER
140  PRINT "#";ENTRY; : INPUT AMOUNT(ENTRY) :
IF AMOUNT(ENTRY) > MAX THEN MAX = AMOUNT(ENTRY)
: REM  FIND THE BIGGEST NUMBER
150  SUM = SUM + AMOUNT(ENTRY)
160  NEXT ENTRY
170  AVERAGE = SUM / NUMBER
180  PRINT : PRINT "WHAT TITLE DO YOU WANT AT
THE BOTTOM?" : INPUT TITLE$ : IF  LEN (TITLE$)
> 38 THEN PRINT "TOO LONG.  ONE LINE ONLY":
HOME : GOTO 180
190  RETURN
200  REM  GRAPH THE DATA
210  ADJUSTFACTR = (YHIGH - YLOW - 1) / MAX
220  SPACING = 1 : IF NUMBER < (XHIGH / 2) THEN
SPACING = 2
225  IF NUMBER < (XHIGH / 3) THEN SPACING = 3
227  IF NUMBER < (XHIGH / 4) THEN SPACING = 4
230  REM  DRAW THE WHITE FRAME FIRST
240  GR : COLOR = WHITE
250  HLIN XLOW - 1,XHIGH + 1 AT YLOW - 1 : HLIN
XLOW - 1,XHIGH + 1 AT YHIGH + 1 : VLIN YLOW -
1,YHIGH + 1 AT XLOW - 1 : VLIN YLOW - 1,YHIGH +
1 AT XHIGH + 1
260  REM  NOW THE BARS THEMSELVES
270  FOR ENTRY = 1 TO NUMBER
280  OVER = OVER + SPACING : COLOR = OVER : IF
INT (OVER / 16) = (OVER / 16) THEN  COLOR = 9 :
REM  NUMBERS EVENLY DIVISIBLE BY 16 WILL BE
BLACK, SO SUBSTITUTE ORANGE
```

```
290  VLIN YHIGH,YHIGH - (AMOUNT(ENTRY) *
ADJUSTFACTR) AT OVER
300  NEXT ENTRY
310  COLOR = WHITE : HLIN XLOW + SPACING,XHIGH
- SPACING AT (YHIGH - (AVERAGE * ADJUSTFACTR))
320  HOWFAROVER = (40 -  LEN (TITLE$)) / 2 :
VTAB 22 : HTAB HOWFAROVER : PRINT TITLE$
330  ROUNDEDAVERAGE = ( INT ((AVERAGE + 0.05) *
10)) / 10 : VTAB 23 : HTAB 10 : PRINT "AVERAGE
= ";ROUNDEDAVERAGE
340  VTAB 23 : HTAB 9 : GET KEYPRESS$
350  RETURN
```

For the Apple, low resolution ("lo-res") gives us a "window" made up of bricklike picture elements (pixels) with 40 across and 40 down. If we make a white border around the whole thing, our graphing field will be 38 by 38. If our vertical distance is only 38, how can we show numbers bigger than 38? What if the number were 76? You might quickly say, "Well, make every brick count as two." What if I said my biggest number was 152? "Make every brick count as four instead." How did you figure that out? It may have just seemed "obvious" to you, but you still have to tell the computer exactly what to do.

Look at the lines of the program (between 100 and 200) where I input the numbers. In line 140, I check to find the biggest number (MAX). How did you figure out that with the biggest number being 76, the bricks had to count for two? You took the biggest number and *divided it by the largest number of pixels you could graph*: 76/38 = 2. If you have a number like 24 on another part of your graph, you will multiply 24 by ½ and get a spot 12 bricks high. I do just that in line 210; YHIGH − YLOW tells how much space I have (38), and then I divide by MAX. Actually, to make the graph work a little better, I subtracted one more from

38, to allow more room. That ADJUSTFACTR is used in line 290 when I make a bar for this graph. I also use it in line 310 to draw a white average line across the bars.

When you run the program, you will see a nice self-adjusting bar graph with a white average line and the value of the average placed at the bottom, along with your title. If you have a color monitor or TV hooked up to your computer, you will see that no two adjacent bars are the same color. That's so that when you do 38 bars, with the lines right next to each other, you can tell the difference.

A HIGH-RESOLUTION GRAPHER

The lo-res program works well as long as you don't get more than 38 points for your graph. What if you want to do more, or want to do a line graph, instead of a bar graph?

Because of the way I wrote it, using variables such as XHIGH and XLOW and ALLOWED, we can easily change this to do a high-resolution ("hi-res") graph. We have to remove two lines, add one new one, and change eight others. Here are the changes:

```
10  REM  HI-RES DATA GRAPHER
15  REM  LINES 225 AND 227 WERE REMOVED; LINE
265 IS NEW
20  XHIGH = 277 : YHIGH = 157 : XLOW = 2 : YLOW
= 2 : ALLOWED = 275 : WHITE = 3 : CHARTREUSE =
1
220  SPACING = INT (((XHIGH - XLOW) - NUMBER) /
(NUMBER + 1))
240  HGR : HCOLOR = CHARTREUSE
250  FOR DOT = 1 TO 2 : HPLOT XLOW - DOT,YLOW -
DOT TO XHIGH + DOT,YLOW - DOT TO XHIGH +
DOT,YHIGH + DOT TO XLOW - DOT,YHIGH + DOT TO
XLOW - DOT,YLOW - DOT : NEXT DOT
```

```
265  HCOLOR = WHITE : HPLOT XLOW,YHIGH -
(AMOUNT(1))
270  FOR ENTRY = 2 TO NUMBER
280  OVER = XLOW + ((ENTRY - 1) * SPACING)
290  HPLOT TO OVER,(YHIGH - (AMOUNT(ENTRY) *
ADJUSTFACTR))
310  POKE 228,29 : HPLOT XLOW + 3,(YHIGH -
(AVERAGE * ADJUSTFACTR)) TO XHIGH - 3,(YHIGH -
(AVERAGE * ADJUSTFACTR))
```

Putting in those same numbers for the car licenses, I get the graph in Figure 2.

It looks okay, but it would be nicer if we could add some numbers down the side so that we could tell what the numbers were, just by looking across from the points to the side.

The Apple has no easy way to put text on the high-resolution screen. On the other hand, the Apple lets you define shapes and then move them anywhere you want. You can even make the shapes larger or rotate them.

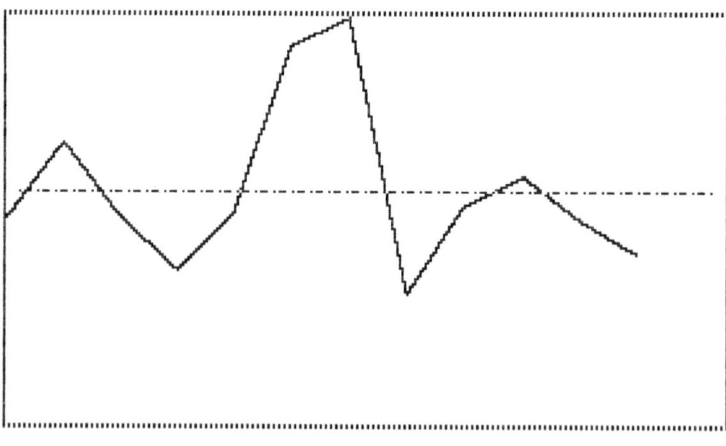

Figure 2. High-resolution graph generated by program given in text

I can define ten shapes that will look like the numbers 0 to 9 and use those. How I came up with those shapes is rather lengthy, so I won't explain it here. Take a look at the bibliography ("Selected Reading") for some books that explain the procedure. To add a module for putting numbers on the screen we need to first put in a "shape table," and then a module in our program to figure out a scale for the sides and put numbers by them. I'm afraid that to do all that takes another twenty-six lines. Here they are:

```
10  REM  MODIFICATIONS TO HI-RES GRAPHER
15  GOSUB 10000 : REM  PUT IN THE NUMBER SHAPE
TABLE FIRST
25  XLOW = 40 : ALLOWED = 235
255  GOSUB 400 : REM  PUT IN THE SCALING LINES
400  REM  SCALE THE SIDES
410  LET MAX$ = STR$ (MAX) : LMAX = LEN (MAX$)
420  INC = 10 ^ (LMAX - 1)
430  START = (INT (MAX / INC)) * INC
440  FOR LINE = START TO 0 STEP - INC
450  IF LINE < 1 THEN 480
460  X = 3 : Y = (YHIGH - (LINE * ADJUSTFACTR))
: N = LINE : GOSUB 900
470  HPLOT XLOW,Y TO XHIGH,Y
480  NEXT LINE
490  RETURN
900  REM  DRAW ONE NUMBER ON THE GRAPH
910  HCOLOR = WHITE : SCALE = 1 : ROT = 1 : N$
= STR$ (N) : FOR PIECE = 1 TO  LEN (N$)
920  DRAW VAL (MID$ (N$,PIECE,1)) + 1 AT (X -
7) + (PIECE * 7),Y : NEXT PIECE : RETURN
10000  REM  HI-RES NUMBERS INITIALIZATION
ROUTINE
10001  HIMEM : 37888
10005  POKE 232,0 : POKE 233,148 : FOR LOC =
37888 TO 38017 : READ QUANTITY : POKE
LOC,QUANTITY : NEXT : RETURN
10009  REM  DATA STATEMENTS OF NUMBER SHAPES TO
```

```
BE POKED INTO A DATA TABLE THAT STARTS AT 37888
10010  DATA 10,0,22,0,36,0,44,0,54,0,66,0,76,
0,88,0,99,0,108,0,119,0
10015  DATA 12,37,28,63,23,54,46,30,14,45,5,
36,4,0,36,188,150,18,45,28,36,0
10020  DATA 101,228,63,23,150,241,46,45,37,0,
12,12,60,63,183,146,21,45,12,228,7,0
10025  DATA 58,39,12,12,12,54,174,55,62,0,56,
39,44,45,245,170,54,23,63,28,4,0
10030  DATA 117,246,63,28,36,229,12,12,45,6,0,
12,12,60,63,183,82,30,46,0
10035  DATA 231,100,45,21,246,14,246,63,28,36,
0,231,100,45,21,54,119,30,30,63,4,0
```

The routine works pretty well, using a method to figure out when to count down the side by 10's, 100's, or whatever. Occasionally, the algorithm for this can get fooled, so a better method is to allow the person running it to adjust the line increments. It takes two more lines:

```
75 IF KEYPRESS$ = "R" THEN PRINT : PRINT "CURRENT LINE INCREMENT = "; INC : PRINT "NEW VALUE - "; : INPUT NINC : HOME : GOSUB 200

425 IF NINC > 0 THEN INC = NINC
```

The final hi-res graph is shown in Figure 3.

WHAT TYPE OF GRAPH SHOULD YOU USE?

A *line graph* is useful when you have a lot of data points that vary over time or some other variable and you wish to show trends in how the data is going. A bar chart is better used when you have just a few quantities that you wish to compare. This can also be

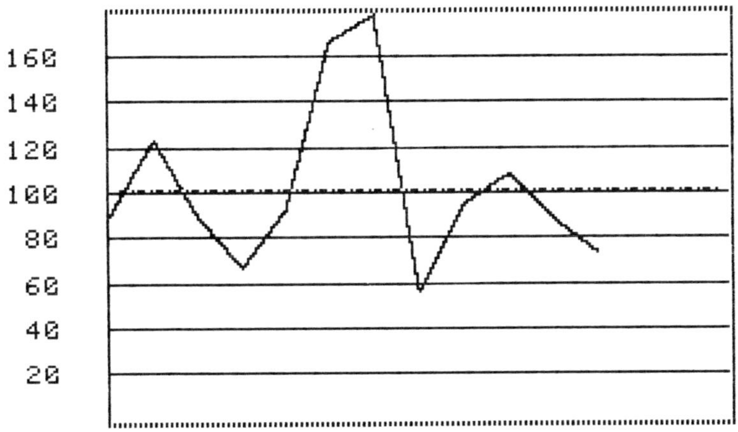

Figure 3. High-resolution graph generated by modified program given in text. On left are numbers of cars licensed. Average is shown by dotted line in the middle.

a time variable chart. A horizontal bar chart compares quantities for all the same variable, say, for an instant of time. You should organize these bars in alphabetical or quantity order to make the graph most usable.

A *pie chart* shows the fractions of a whole as slices out of a pie. It also shows one instant of time, but makes it easier to see how much of the whole one segment makes up. A good idea is to put the largest slice of the pie in the upper section where the eye naturally goes first.

A variation on the line graph is the *area graph*. In this one, there are several line graphs, but each one is filled in with color or a pattern down to the bottom. Put your darkest color or pattern on the lowest, most stable, or most important quantity. Area charts are easier to interpret than putting several lines alone on one graph.

The *3-D bar graph* is impressive looking but won't

necessarily give you better information. Using it in the two-row form, as in comparing growth rates over a year for two different types of fertilizing, will always make the information in the back row look larger, just by perspective. You can make your graphs show just about whatever you want. By using a small increment scale, you can emphasize changes, if that is what you want to show. By using large increments, you can minimize changes if your project would be better served by that.

COMMERCIAL SOFTWARE

Although the two graphing programs you typed in above are fairly well done, you can also buy commercial programs to do a better job or do some of those other kinds of graphs. Some of the commercial programs can give you a help with the mathematics by analyzing the data to see if your points could lie on a particular line. These "curve fitters" will try to come up with a mathematical function that, when plotted, will closely match your points.

Many of these programs are available, ranging in price from $24.95 to several hundred dollars. If you or your school want to buy such a program, make sure the dealer allows you to try out a copy at the store first. Check for the following points:

1. Is the software easy to use?
2. Will it do the types of graphs you really want?
3. Can you save data on disk or tape so you don't have to retype to try something different?
4. Are there curve-fitting programs to go with the software (assuming you want that sophistication)?

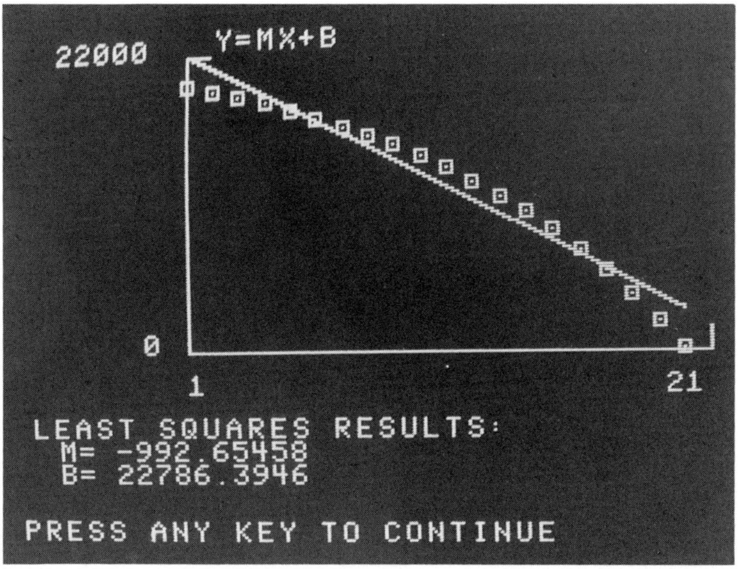

NCSSM's Tool *has generated this curve. The "straight" line is the best fit of an equation of the type Y = MX + B, using the values at the bottom of the screen.*

5. Can you print out the graphs on your printer?
6. Is the price low enough so that you or one of your friends couldn't do a better job yourselves?
7. Does the software come with a backup disk or ways to make one?
8. Does the manual give you enough information and suggestions so that you are eager to do things you never thought of before?

I can give some specific recommendations also: For Atari and Commodore 64, *B/Graph* is impressive in the charts and statistics that it will do. Prices ranged from inexpensive to moderately so ($40–$100), depending on the version and discount. Radio Shack presents a problem. The TRS-80 line has been discontinued, and most stores will not have much current since they are moving into IBM-compatible programs. There are very good business graphing programs for IBM machines, but they tend to be fairly expensive. TRS-80 users' groups, however, can probably give you some good ideas and even some public-domain programs that will work reasonably well.

For the Apple, quite a large number of graphing programs are available, some of the best not in the usual retail stores. For example, *Graphic Analysis II* (Vernier Software) does nice graphing and good statistics, and is quite inexpensive. *Curve Plotter!* and *Alpha Chart* (Spectral Graphics Software) do line, area, bar, stacked-bar, and 3-D bar graphs; do best-fit graphics; have a graphics editor; and are quite inexpensive.

In the higher price range, you might investigate *pfs:graph* and *The Graphics Department* (Sensible Software). The latter program also does a host of other functions related to graphics, slides, and business presentations.

IDEAS FOR SCIENCE PROJECTS

What sorts of things can be graphed in a project? For one thing, in the final presentation to a teacher, science fair, or other students, a good graph can really dress up your report, besides making things clearer.

Moreover, there are projects where graphing is done at the start to give you ideas for further investigation. The parking lot study is an example of a time-dependent study. Another might be for you to go to your county or city hall of records and do some research into populations. What months do people seem to die in; is there some sort of pattern? Has this changed since the last century? Can you correlate advances in medicine with this or is there another explanation that might fit better?

Check the local hospital records (not names or details, just total admissions) to see if more people are admitted during certain times of the year. Can you find out if these are for the emergency room (accidents primarily) or for other types? Does "elective surgery" have a particular "season"?

What about finding out the total population of the county for each decade since the records were started (the United States census is taken every decade). Then you can take the number of deaths per thousand population over the last hundred years or so. Has there been a change? Has the change been steady? If it is not a steady decrease, can you figure out why the peaks and valleys are there? Do wars show up? Does the building of a new hospital fit the date of a special point on the curve? How about a graph of ages at death for different time periods?

Right now, scientists are looking for old air, the type you might find in bottles sealed a hundred years ago, (or in hollow brass buttons!) to see if the carbon dioxide level of the atmosphere has changed in the last century. Why not look for other time-dependent things particular to your locality. If you live in a farming community, has production of certain crops, livestock, or products changed with time? Why? If you can find the records for a city (check the department of commerce or agriculture for a state), has *consumption* of certain things changed? What about consumption of dairy products? How about tobacco products before and after the surgeon general's report? Did your local community do the same as national patterns? If not, can you explain the differences?

Another type of project that graphs will help to clarify are distance-dependent studies. How thick is the topsoil in your area? How thick is it near and far away from the nearest river? How deep do wells have to be drilled as you go away from a river?

If you can do a chemical test, you might want to graph how much lead is found on grass along a highway (presumably from leaded gasoline). Do you see cattle feeding on that grass in your area? How far away from the highway should they be kept to avoid lead contamination of meat or milk?

Use some pH paper to tell the acidity or alkalinity of streams, ponds, and lakes in your area. Does a distance-dependent factor show up? Is there a difference depending on how far away the water is from a mill, coal-burning power station, sewage treatment plant, landfill, or highway?

In all these projects, being able to show a graph of your data will make for not only a neater presentation but also one that communicates the information better. Both you and the people you try to communicate your findings to can then see some of the patterns in the scientific world.

3

STATISTICS

"Statistics" is a word that you hear a lot these days. We have the statistics that show that smoking causes lung cancer, statistics that show that wearing seat belts saves lives, and statistics showing that many people don't believe statistics.

It's true, you have to watch out how some of these are figured. For example, in an article in the newspaper touting the virtues of a high-fiber diet (one with a lot of raw vegetables and bran and not so much meat and fats), a comparison was made between the United States and a section of Africa. *According to the statistics*, people on a high-fiber diet in that part of Africa have a much smaller chance of dying from heart disease than do residents of the United States. The conclusion drawn from these "statistics" was that the large amount of fiber in the African diets "protected" these people from heart disease.

However, you have to be careful how you calculate these things. Even though the conclusions could be reached from their data, the problem was the "sample populations." If you limited the grouping to only those people over the age of fifty-five, it turned out that the trend was *reversed*. People over fifty-five in that part of Africa had a much greater chance of dying of heart disease than did people in the same age group in the United States.

You see, those people had a very *good* chance of dying of things like cholera, typhoid, tuberculosis, malaria, snakebite, and malnutrition long before they ever reached the age of fifty-five. In the United States, hardly anybody dies of those things anymore and so a lot more people live long enough to get the degenerative diseases like heart problems and cancer.

I could cite many examples. Just be warned that you need to be careful how you group your samples.

WHY USE STATISTICS?

If statistics cause so much trouble, why are they used in science?

Scientists are lazy.

Perhaps you can identify with that. Suppose you are trying to get some of the "stats" on a football team and you have the weights of the linemen and their substitutes:

185 230 130 178 267 137 187 177 201 212 pounds.

There are only ten of these, but even ten three-digit numbers are hard to remember. One easy way to get a feeling for what the team would be like would be for you to take the **average**. You add up all ten weights and divide by the number of players. I get 1,904 pounds divided by ten players, or 190 pounds (190.4 rounded off) as the weight of the average player.

"AVERAGE" VERSUS "NORMAL"

Does this mean that a normal football lineman weighs 190 pounds? No, average and normal are not the same thing. Think about the projection that the average family in the United States soon will have 1.3 children. Will there be any *average* families in the U.S.? I hope not. Having a third of a child around tends to be rather messy and gruesome. On the other hand, a *normal* family will have three, two, one, or even no children in it.

How do you determine what's normal when you're dealing with something totally new to you? There are ways of doing that using statistics.

First, let's consider what normal does mean. Take a look at Figure 4. This graph has been called a bell-shaped curve, mainly because the curve does look a lot like the traditional bell shape.

The height of the graph tells how many there are of something; the bottom is the variable. For example, I could put skin color of humans of the world on that graph with pure black on the left and pure white on the right. Most people would fall somewhere in the middle with a sort of brown skin, making the hump in the center represent the majority or "normal." But skin color is the result of many factors, including race, historical climate, exposure to sun, environmental factors, and genetic accident. How can all of these even out to such a nice curve? Let me show you with a computer simulation of a similar situation.

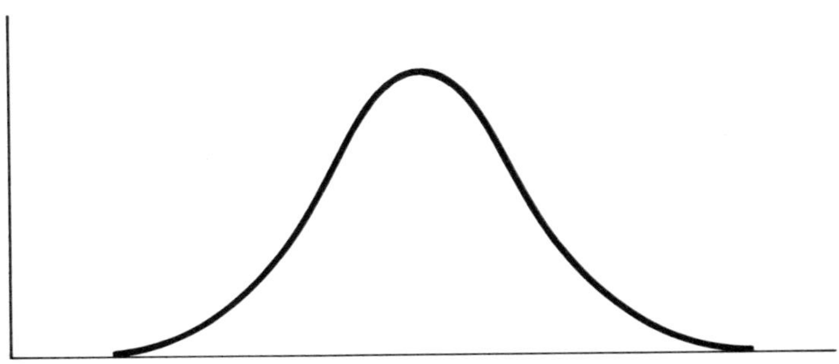

Figure 4. Bell curve

A COMPUTER SIMULATION OF "NORMAL"

When I was in junior high, I made a device that drove my parents insane. That wasn't my intention, but it had that effect, just from the noise. I got a piece of plywood and put nails and wires in it like in Figure 5.

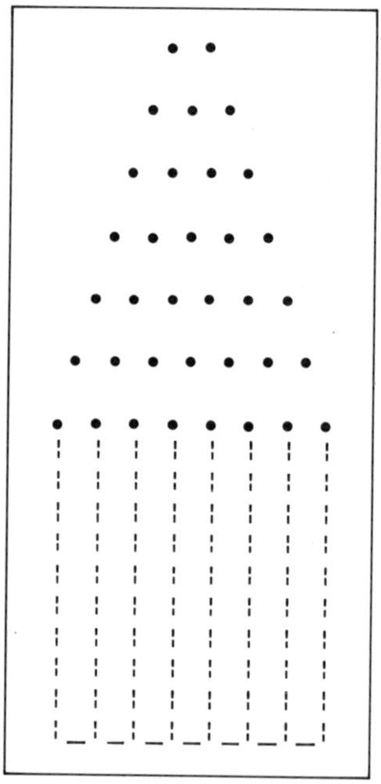

Figure 5. Apparatus for measuring distribution of marbles

After stringing the wires in channels on the bottom, I started dropping marbles between the top two nails and letting them plink-plink their way to the bottom traps. I did it for hours. The thing that intrigued me was that no matter how many times I did it I always got the same pattern of marbles filling up the rows in the bottom. It was a bell-shaped curve. Each marble had to randomly "make" six yes/no "decisions." As it hit each nail directly below it, there was a fifty-fifty chance that it could go to the left or the right. Then it would hit the next nail below that and again have the same left or right chance.

You can make this device if you wish. It should work, although I don't know if your parents will like it any better than mine did. You'll learn the same things with less work by typing in the simulation program below.

```
10  REM  BELL-SHAPED-CURVE GENERATOR
20  BLACK = 0 : RED = 1 : NEUTRAL = 5 : WHITE =
15 : KEYBOARD =  - 6384 : KYRESET =  -16368 :
SPEAKERCLICK =  -16336 : YMAX = 159 : XMAX =
279 : HWHITE = 7
30  BUZZ = 10 : PAUSE = 100
35  DIM XHIRES(10),COUNT(20)
40  FOR PLACE = 1 TO 9 : XHIRES(PLACE) = (31 *
PLACE) - 15 : NEXT PLACE : REM  SET UP GRAPH
SPACING
50  HOME
60  GOTO 500
300  REM  DRAW THE BALL
310  COLOR = RED : PLOT XSPOT,YSPOT : RETURN
400  REM  UNDRAW THE BALL
410  COLOR = BLACK : PLOT XSPOT,YSPOT : RETURN
500  GOSUB 1000 : REM  DRAW THE FIELD
600  GOSUB 2000 : REM  START DROPPING BALLS
700  GOTO 500 : REM  START DOING IT AGAIN AFTER
SEEING THE GRAPH
800  END
1000  REM  DRAW THE FIELD
1010  GR : REM  SET LO-RES GRAPHICS MODE
1020  STARTROW = 11 : FINISHROW = 27 : OFFSET =
1 : COLOR = WHITE
1025  REM  DRAW THE WHITE TARGETS AT THE BOTTOM
1030  FOR OVER = STARTROW TO FINISHROW STEP 2
1040  PLOT OVER,28
```

```
1050  NEXT OVER
1100  REM  NOW DRAW IN THE PEGS
1120  COLOR = NEUTRAL
1130  FOR DOWN = 24 TO 10 STEP - 2
1140  STARTROW = STARTROW + OFFSET : FINISHROW
= FINISHROW - OFFSET
1150  FOR OVER = STARTROW TO FINISHROW STEP 2
1160  PLOT OVER,DOWN
1170  NEXT OVER
1180  NEXT DOWN
1200  RETURN
2000  REM  START DROPPING THE BALLS
2010  XSPOT = 19 : YSPOT = 5 : DROP = 1 : REM
PLACE BALL STARTS TO FALL HOW FAR TO DROP
2020  GOSUB 300 : REM  DRAW THE BALL
2030  FOR TIME = 1 TO PAUSE : NEXT TIME
2040  IF SCRN(XSPOT,YSPOT + DROP) <  > NEUTRAL
THEN 2200
2050  LET CHANCE =  RND (1) - .5
2060  LET MOVE = SGN(CHANCE) * 1
2070  LET SOUND = PEEK (SPEAKERCLICK) : SOUND =
PEEK (SPEAKERCLICK)
2080  LOOK = PEEK (KEYBOARD) : IF LOOK > 128
THEN GOSUB 3000 : GOTO 2400 : REM   GRAPH
RESULTS SO FAR, THEN RETURN TO START NEW BALL
DROPPING
2090  GOSUB 400 : REM  UNDRAW THE BALL AT ITS
CURRENT POSITION
2100  XSPOT = XSPOT + MOVE : GOSUB 300 : REM
DRAW BALL IN NEW POSITION
2110  GOTO 2020
2200  IF SCRN(XSPOT,YSPOT + DROP) <  > WHITE
THEN 2300
2210  LET COUNT(XSPOT - 10) = COUNT(XSPOT - 10)
+ 1 : REM  INCREASE TOTAL FOR TARGET HIT BY ONE
2220  GOSUB 400 : REM  UNDRAW
2230  VTAB 23 : FOR TARGIT = 1 TO 17 STEP 2 :
VTAB 23 : HTAB (TARGIT * 2.3) : PRINT
COUNT(TARGIT) : NEXT TARGIT: REM  DISPLAY
CURRENT TOTALS AT BOTTOM
```

```
2240  FOR CLICK = 1 TO BUZZ : SOUND =  PEEK
(SPEAKERCLICK) : NEXT CLICK
2250  GOTO 2000 : REM  DROP NEXT BALL
2300  GOSUB 400 : REM  UNDRAW BALL IF NOTHING
GOING TO BE HIT
2310  YSPOT = YSPOT + DROP
2320  GOSUB 300 : REM  DRAW IT IN NEW POSITION
2330  GOTO 2040 : REM  CHECK TO SEE WHAT WILL
HAPPEN NEXT
2400  RETURN
3000  REM  HI-RES GRAPH OF CURRENT TARGET
TOTALS
3005  POKE KYRESET,0 : REM  RESET KEYBOARD
READER FOR NEXT TIME
3010  HGR : HCOLOR = HWHITE : REM  SET HI-RES
GRAPHICS MODE, CHOOSE COLOR
3030  XLOC = 0
3040  FOR TARGIT = 1 TO 17 STEP 2
3050  XLOC = XLOC + 1 : HPLOT XHIRES(XLOC),YMAX
TO XHIRES(XLOC),YMAX - (COUNT(TARGIT))
3060  NEXT TARGIT
3070  XLOC = 1 : HPLOT XHIRES(XLOC),YMAX -
(COUNT(1))
3080  FOR LINE = 3 TO 17 STEP 2
3090  XLOC = XLOC + 1
3100  HPLOT TO XHIRES(XLOC),YMAX -
(COUNT(LINE))
3110  NEXT LINE
3200  VTAB 24 : PRINT "<PRESS ANY KEY>"; : GET
KP$ : PRINT : HOME
3210  RETURN
```

The program simulates the dropping marble, using graphics. In low-resolution graphics on the Apple, a red "ball" bounces its way down seven rows of gray "nails," hitting one of nine white "targets" at the bottom. Instead of marbles piling up in channels, an array of locations in the computer memory keeps track of how many times each target has been hit.

If you know some BASIC programming, you might try to modify the program to use a different number of nails. If you've done very little programming so far, don't try to understand my explanation. I'll just touch on the highlights. The rest of the program should be easy to understand for those of you who've worked with a computer a while. The use of the SCRN and RND functions here might give you some ideas for making up your own games.

The main logic of the program uses the SCRN function. Before moving the ball to the next point lower, I look at that point using the SCRN function to tell me what color the next spot already is, as in lines 2040 and 2200. If it's black, that means nothing's there and the ball drops (2310, 2320). If the next spot is gray (NEUTRAL), that means there's a nail in the way and I have the computer use the RND function to randomly pick a way the ball will go (2050, 2060). If the next spot is white, I'll be hitting a target and so the count for that target is increased by one (2210). Then the next ball is dropped.

Anytime a key is pressed, a high-resolution line graph of the current count is drawn. Pressing another key continues the count. I've even included a sound routine (2070, 2240) so that you can annoy others.

As you run the program, the graph builds up slowly. The longer you run it, the more the curve plotted begins to look like the bell-shaped curve I drew above. Figure 6 shows what it looked like when 144 balls had dropped into the middle channel.

By the way, if you want grades or some other small number to fall into a curve like this, you have to use regions, such as scores of 0-10, 11-20, or 21-30, or you won't get enough repeats to make any hump.

With a bell-shaped curve, the big hump in the middle includes most of the examples. If you've heard of "grading on the curve" for a class, this is the curve

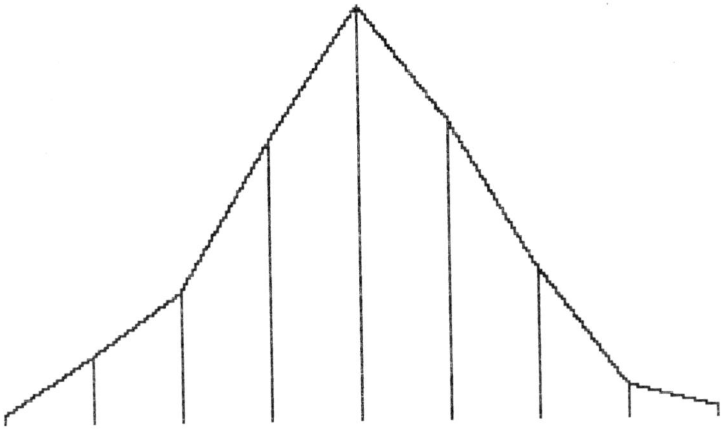

Figure 6. Computer-generated "bell curve"

they were talking about. The C's would be the big hump in the center, the B's the right shoulder, and just a few A's would be on the far right. How do you know where the hump begins and ends without graphing things every time? A statistical technique called the "standard deviation" will do it for you.

AVERAGE AND "AVERAGE DIFFERENCE"

You have to sneak up on this one for most people or it won't make any sense. First, let's take the series of numbers we got for the weights of the football players:

185 230 130 178 267 137 187 177 201 212 pounds

We found the average weight to be 190.4 pounds by dividing the total weight by the number of players. Then we can ask, "How much does each kid differ from the average, since nobody is exactly at the average?" Let's set up Table 1.

Table 1

Kid No.	Weight	Average	(Weight) - (Average) Difference from Average
1	185	190.4	− 5.4
2	230	190.4	39.6
3	130	190.4	−60.4
4	178	190.4	−12.4
5	267	190.4	76.6
6	137	190.4	−53.4
7	187	190.4	− 3.4
8	177	190.4	−13.4
9	201	190.4	10.6
10	212	190.4	21.6

Now what is the *average difference*—about how much is the usual kid away from the regular average? To do that, you have to ignore the plus and minus signs to add up the numbers. I got 296.8 divided by 10, or 29.68. But remember, we ignored the positive and negative signs. We should then call that "plus or minus 29.68 pounds" as our average difference. We could then fairly confidently say that the average kid in our group had a weight of 190.4 pounds, + or − 29.68 pounds. No kid had *exactly* 190.4 pounds for a weight, but most of them fell into the range of about 160 to 230 pounds. Players number 3, 5, and 6 didn't, but most of the kids did.

VARIANCE AND STANDARD DEVIATION

Figuring out the average difference this way is not too unreasonable, and the answers you get seem to make sense. How is this different from the standard deviation I spoke about above? The difference comes from the mathematical trick you use to figure about the same thing.

Remember that I told you above to ignore the plus and minus signs? One way to do this is to get rid of them by multiplying each number by itself—squaring it. Perhaps you have learned by now that multiplying a negative number by another negative number results in a positive number. Of course, multiplying two positives also gives a positive result. This is the way we can get rid of all the negatives. Let me add to that chart. The result is Table 2.

Table 2

Kid #	Weight	Average	Difference	Diff^2
1	185	190.4	− 5.4	29.16
2	230	190.4	39.6	1568.16
3	130	190.4	−60.4	3648.16
4	178	190.4	−12.4	153.76
5	267	190.4	76.6	5867.56
6	137	190.4	−53.4	2851.56
7	187	190.4	− 3.4	11.56
8	177	190.4	−13.4	179.56
9	201	190.4	10.6	112.36
10	212	190.4	21.6	466.56

For that column of squared differences, I got a total of 14,888.4 or an average of 1488.84. This is called the **variance.** Of course, for our purposes, to get a number we can identify with, we have to undo all that squaring we did. Taking the square root is easy with a calculator or a computer. I got 38.59. This is the **standard deviation.**

The neat thing about using the standard deviation is that if you get a large enough group of samples the large hump that contains a majority of the examples will be within one standard deviation above or below the average. If you mark everything + or − one standard deviation, you should get 68% of all the examples. If you go *two standard deviations* to either side of the average, you ought to get 95% of everything. Using the standard deviation can tell you where the sides of the hump should be, without having to graph it. If something falls outside of the main group, you know it's got to be pretty unusual and suitable for more study, such as in a science project.

MAKING THE COMPUTER DO THE WORK

Now that you know how to do it, you can see that actually calculating the average, variance, and standard deviation can be repetitive, messy, and boring. This is exactly the sort of things computers do best. The program below should do the trick, making a chart much like Table 2.

```
10  REM  AVERAGE AND STANDARD DEVIATION
20  D$ = CHR$(4) : SUM = 0 : SQUARESUM = 0 :
DSUM = 0 : DECIMALFLAG = 0
50  GOSUB 100 : REM  GET THE DATA
60  GOSUB 200 : REM  DO CALCULATIONS AND PRINT
IT OUT
70  END
100  REM  GET THE DATA
110  HOME : PRINT "How many numbers are you
going to be" : PRINT "typing in"; : INPUT
NUMBER
120  DIM AMOUNT(NUMBER + 1) : PRINT
130  FOR ENTRY = 1 TO NUMBER
140  INPUT AMOUNT(ENTRY) : IF INT
(AMOUNT(ENTRY)) <  > AMOUNT(ENTRY) THEN
DECIMALFLAG = 1
150  NEXT ENTRY
```

```
160  RETURN
200  REM  CALCULATIONS AND PRINT OUT
210  HOME : PRINT "Do you want this to go to
the (S)creen" : PRINT TAB( 31);"(P)rinter"; :
GET PO$ : PRINT PO$ : IF PO$ <  > "S" AND PO$ <
> "P" THEN 210
220  FOR ENTRY = 1 TO NUMBER : SUM = SUM +
AMOUNT(ENTRY) : NEXT ENTRY
230  AVERAGE = SUM / NUMBER
240  IF PO$ = "P" THEN  PRINT D$;"PR#1" : REM
TURN ON THE PRINTER IN SLOT 1
250  PRINT "ENTRY"; SPC( 5)"AMOUNT"; SPC(
4);"DIFFRNCE"; SPC( 2);"DIFF.SQRD"
260  FOR HYPHEN = 1 TO 40 : PRINT "-"; : NEXT
HYPHEN : PRINT
270  FOR ENTRY = 1 TO NUMBER
280  PLACESNEEDED = 3 : N$ = STR$(ENTRY) :
GOSUB 450 : ENTRY$ = N$ : PLACESNEEDED = 10 : N
= AMOUNT(ENTRY) : GOSUB 400 : AMOUNT$ = N$ :
DYFFERENCE = AMOUNT(ENTRY) - AVERAGE : N =
DYFFERENCE : GOSUB 420 : DYFFERENCE$ = N$
290 DSUM = DSUM +  ABS (DYFFERENCE) :
SDYFFERENCE = DYFFERENCE ^ 2 : SQUARESUM =
SQUARESUM + SDYFFERENCE : N = SDYFFERENCE :
GOSUB 420 : SDYFFERENCE$ = N$
300  PRINT ENTRY$; SPC( 3);AMOUNT$; SPC(
2);DYFFERENCE$; SPC( 2);SDYFFERENCE$
310  NEXT ENTRY
320 N = AVERAGE : GOSUB 420 : AVERAGE$ = N$ :
ADYFFERENCE = DSUM / NUMBER : N = ADYFFERENCE :
GOSUB 420 : ADYFFERENCE$ = N$
330 VARIANCE = SQUARESUM / NUMBER : N =
VARIANCE : GOSUB 420 : VARIANCE$ = N$:STDD =
SQR(VARIANCE) : N = STDD : GOSUB 420 : STDD$ =
N$
340  FOR HYPHEN = 1 TO 40 : PRINT "-"; : NEXT
HYPHEN : PRINT
350  PRINT "AVERAGE:"; SPC( 11);AVERAGE$ :
PRINT "AVERAGE DIFFERENCE:";ADYFF$ : PRINT
"VARIANCE:"; SPC( 10);VARIANCE$ : PRINT "STD.
```

```
DEVIATION : +/-";STDD$
360  IF PO$ = "P" THEN  PRINT D$;"PR#0"
370  RETURN
400  REM  NUMBER FORMATTING
410  IF DECIMALFLAG = 0 THEN N$ = STR$ (N) :
GOTO 450
420  N = (INT ((N + 0.005) * 100)) / 100 : N$ =
STR$ (N) : REM  ROUND OFF TO TWO DECIMAL PLACES
AND CONVERT TO STRING
430  IF INT (N) = N THEN N$ = N$ + ".00"
440  IF MID$ (N$,( LEN (N$) - 1),1) = "." THEN
N$ = N$ + "0"
450  LN = LEN (N$) : IF LN < PLACESNEEDED THEN
N$ = " " + N$ : GOTO 450
460  RETURN
```

When you get the program typed in, test it using the weights of the players from the table above. You should get the table with four extra lines at the bottom that look like this:

AVERAGE:	190.40
AVERAGE DIFFERENCE:	29.68
VARIANCE:	1488.84
STD. DEVIATION: +/−	38.59

SPREADSHEET PROGRAMS

Back in the "ancient history" in the microcomputer world—about 1978—two men developed a program called *VisiCalc*. This was the first successful **spreadsheet** program that mimicked something the business community had been using for years.

Spreadsheets used to be big sheets of paper marked off in a grid for holding information. For example, department stores had been using the idea for one sheet that would put all the sales for each department down one side and then do that for each day of a week as you went across in a grid. At the end of the week, adding up all the columns in a row across would give you the sales for that department. Adding a column down would get you the total sales for all departments for that day. From there, a manager could add other columns to get other useful figures, such as sales per square foot of space for each department or cost for salaries per sale. The information was great for planning, but the calculations involved were tremendous. Then came the computer spreadsheet, and the microcomputer industry really took off.

You can use a spreadsheet too. Look at the statistical program output in Table 3 (just part of the list and modified a little).

Table 3

	A	B	C	D	E
1	Kid #	Weight	Average	Difference	Diff^2
2					
3	1	185	190.4	−5.4	29.16
4	2	230	190.4	39.6	1568.16
5	3	130	190.4	−60.4	3648.16
6	4	178	190.4	−12.4	153.76
7					

Instead of writing a special program to make up this sheet with columns of numbers spread out on it, we could have told a spreadsheet program to do the same thing, saving many hours of programming. In such a program, you can put labels in each cell, such as "Kid No." in cell A1 and "Weight" in cell B1. In cell D, I would put the formula (B–C), that is, subtract the value in C from that in B. The formula for the cells in column E would be (D*D), meaning D times D.

After you enter the data, you can say "recalculate"

and all the numbers in the other columns will be recalculated to reflect the newest data and the screen display will change as you watch. How fast this happens depends on how good the program is and, unfortunately, usually on how much it costs.

The original makers of *VisiCalc* are now out of business, but your school probably has a copy of that program or *AppleWorks*. Both are fairly expensive, but cheaper and sometimes better substitutes are available. There are a tremendous number of spreadsheet programs for the Apple, Commodore 64, Atari, and all other machines, including the original *VisiCalc* on many of them. Just look for the suffix "calc". Check your local computer club for the machine you have access to and see if they have an inexpensive spreadsheet.

For many science projects that involve a large number of repetitive calculations, using a spreadsheet program may be much easier than finding a commercial science program or writing one yourself.

HOW TO USE STATISTICS IN A SCIENCE PROJECT

Standard deviation and other statistical techniques can make your science project much better.

If your hypothesis for a project is that the angle of light on a plant is important to its growth, you could set up a lot of plants with the light angle controlled by placement of light sources. At the end of some time, you would probably see some differences among the various plants subjected to differing conditions.

A key question used in experiments of this type, however, is, "Are the differences *statistically significant?*"

If you took the average height of the plants with light overhead and compared them with the average height of plants with the light at a low angle, you would probably find a difference. However, if, after using the statistics program above, you found one of the averages to be within one standard deviation of the other, the results would not be statistically significant. To a scientist, your hypothesis would not have been proven. What seemed to be differences are within the normal variability of that plant species. Some researchers require that the differences be *two* standard deviations away from the average. Since 95% of the normal individuals should be within that area, they can then say that they are "95% confident" that they have discovered a significant difference. There is only a 5% chance the differences could have happened without the changes you made.

There is much more to statistics than this, but you now have one of the fundamental methods used. If you want to go further, the next topics you should look up in a statistics book are the "one-tail test" and the "two-tail test." They involve just a little more math but can help to justify significance in results which are not very different from the control case. See Appendix 1 for more on statistics.

IDEAS FOR SCIENCE PROJECTS

Statistics are needed in many science projects where you get a large amount of data. By using the proper calculations, you can tell if your results really say your experiment proved anything significant, or just showed small variations.

Biology projects are usually easier to do than chemistry or physics ones. You can take almost any plant, change some conditions and you will get a change in its growth pattern. For animal experiments, there are rigid controls on the use of any animal with a backbone so you have to be very careful in experimental design.

What about variations on the plant experiment? You could perhaps use as a variable sunlight versus indoor light for the same amount of time. You could change the type of indoor light, such as fluorescent or regular tungsten (incandescent) light bulbs. You might wish to try the different types of fluorescent lights: the standard cool white, the new daylight bulbs, or even the purplish "grow lights."

Other variations on your plant experiment could be in how you water the plants: daily just a little bit, or once a week and a lot (using the same total amount of water over the time period), or perhaps you might want to investigate the effect of soil temperature on plant growth.

You might want to ask a chemistry teacher for some hints on looking up the problem of acid rain and see what effects small amounts of very dilute sulfurous acid solutions have on plants. Does it affect all plants the same way? Are some plants more resistant to its effects?

How about the soil around your town—can you vary the amount of nitrogen, potassium (potash), and phosphorus you add to see what levels give the best growth for your type of plant? What about how acid or alkaline the soil is? Does the addition of lime, an alkali, change how your plants grow when mixed in varying amounts with the soil around you?

Atomic and molecular physical chemists use statistics to tell what populations of particles do. They can't tell what any one atom, proton, or molecule will do, but they can tell what *groups* do.

Psychologists and sociologists are faced with the same situation. It's difficult to predict what an individual will do, but groups are easier to handle, if you use statistics.

You might try an experiment in those fields yourself. What happens when (with permission of the principal) you put a group of chairs out in the courtyard? Do they stay in the same place, or do people move them? How often? What type of people move them the most? What if you put up a barricade in a hallway without anything visible for a reason? Do people go around it or avoid that hallway? Do boys do one thing and girls another? Do seniors act one way and sophomores differently? What if you take a survey in the student newspaper about a controversial question. You can get the average response, but are the strange answers within the standard deviation and so not really unusual?

Statistics, when the computer handles the endless calculations, can make it easier to draw conclusions from your data. In turn your project may end up more successful and any science fair presentation will be better organized and more meaningful.

4 SORTS AND DATA BASES

SORTING

There are times when you have gathered a lot of data and perhaps done some preliminary calculations, when you find yourself wanting to put everything into some sort of order.

For example, suppose you have collected all the heights of various trees in your part of the city and you wish to put them in order from the smallest to the largest, so that you can easily see which trees are about the same height. If you have a way to keep track of other facts associated with each tree, you may be able to make a hypothesis as to what is important in the growth patterns. This organizing process is called **sorting.**

HOW A COMPUTER SORTS

Most people alphabetize things, such as library cards, by putting the first two cards in order, taking the next and flipping through until they get to the right spot and inserting the new card. Computers can do it the same way. This method is called the "Insertion Sort," and the computer algorithm is fairly simple. You can watch the computer sort some lines by typing in the program below:

```
10  REM  INSERTION SORT DEMO PROGRAM
20  WHITE = 3 : BLACK = 0
30  GOSUB 100 : REM  PICK THE NUMBER
40  GOSUB 200 : REM  GENERATE SOME RANDOM
AMOUNTS
50  GOSUB 300 : REM  DRAW THE LINES
60  VTAB 22 : PRINT "PRESS ANY KEY TO START"; :
GET KEYPRESS$
70  GOSUB 900 : REM  SORT AND DISPLAY
80  END
100  REM  CHOOSE HOW MANY
110  HOME : PRINT "HOW MANY LINES DO YOU WANT
TO SORT" : PRINT "(5-130)"; : INPUT NUMBER : IF
NUMBER < 5 OR NUMBER > 130 THEN 110
120  DIM AMOUNT(NUMBER + 1)
130  RETURN
200  REM  MAKE UP THAT MANY RANDOM NUMBERS
210  FOR ITEMS = 1 TO NUMBER : AMOUNT(ITEM) =
INT (RND(1) * 100) + 1 : NEXT ITEM
220  RETURN
300  REM  PLOT THE MESS
310  HGR : HCOLOR = WHITE
320  FOR LINE = 1 TO NUMBER : HPLOT 0,LINE TO
AMOUNT(LINE),LINE : NEXT LINE
330  VTAB 21 : PRINT "THE MESS"
340  RETURN
900  REM  INSERTION SORT
910  FOR TIMESTHROUGH = 2 TO NUMBER
920  LET TEMP = AMOUNT(TIMESTHROUGH)
930  FOR PLACE = TIMESTHROUGH - 1 TO 0 STEP -1
940  IF AMOUNT(PLACE) < TEMP THEN 970
950  HCOLOR = BLACK : HPLOT 0,(PLACE + 1) TO
```

```
100,(PLACE + 1) : LET AMOUNT(PLACE + 1) =
AMOUNT(PLACE) : HCOLOR = WHITE : HPLOT 0,(PLACE
+ 1) TO AMOUNT(PLACE + 1),(PLACE + 1)
960  NEXT PLACE
970  HCOLOR = BLACK : HPLOT 0,(PLACE + 1) TO
100,(PLACE + 1) : AMOUNT(PLACE + 1) = TEMP :
HCOLOR = WHITE : HPLOT 0,(PLACE + 1) TO
AMOUNT(PLACE + 1),(PLACE + 1)
980  NEXT TIMETHROUGH
990  RETURN
```

The actual sorting routine is in lines 900–980; the first part of the program just draws the random-length lines to be sorted. Line 920 picks the next line to be looked at and calls it TEMP. The program looks at the list of lines and keeps skipping over the lines that are smaller (line 940). If the proper place is found, line 950 moves everything else down and places the target line into place.

A computer can sort things a lot of other ways. An even simpler sorting method, called the "Bubble Sort," involves only about four program lines. Unfortunately, although simple to understand, it is the slowest way to sort things. The Insertion Sort demo program run on 100 lines took 155 seconds. A similar program using the Bubble Sort took more than twice as long. The Insertion Sort works quite well, especially when you want to add a few items to a list of things that are already sorted.

Another sorting method is called "Quicksort," which was invented by C.A.R. Hoare in 1962. The program looks confusing but is actually faster than anything else. The program below will sort 100 lines in only 26 seconds—over twelve times faster! The advantage of this sort gets much better as you try to sort more things. It's easier to watch what's happening than to explain it. Take the Insertion Sort demo program you just typed in and SAVE it. Now delete lines 900–990 and type in these lines:

```
300  REM  QUICK SORT
305  DIM B(22) : REM  ACTUALLY NEEDS 2LOGN
(BASE2).  22 IS GOOD ENOUGH FOR 2000 ITEMS.
310  L = 1
320  B(L) = NUMBER + 1
330  M = 1
340  J = B(L)
350  I = M - 1
360  IF J - M < 3 THEN 520
370  M1 =  INT ((I + J) / 2)
380  I = I + 1
390  IF I = J THEN 460
400  IF AMOUNT(I) <  = AMOUNT(M1) THEN 380
410  J = J - 1
420  IF I = J THEN 460
430  IF AMOUNT(J) > = AMOUNT(M1) THEN 410
440  HCOLOR= 0 : HPLOT 0,I TO 100,I : HPLOT 0,J
TO 100,J : AMOUNT(0) = AMOUNT(I) : AMOUNT(I) =
AMOUNT(J) : AMOUNT(J) = AMOUNT(0) : HCOLOR = 3
: HPLOT 0,I TO AMOUNT(I),I : HPLOT 0,J TO
AMOUNT(J),J
450  GOTO 380
460  IF I > = M1 THEN I = I - 1
470  IF J = M1 THEN 490
480  HCOLOR = 0 : HPLOT 0,I TO 100,I : HPLOT
0,M1 TO 100,M1 : AMOUNT(0) = AMOUNT(I) :
AMOUNT(I) = AMOUNT(M1) : AMOUNT(M1) = AMOUNT(0)
: HCOLOR = 3 : HPLOT 0,I TO AMOUNT(I),I : HPLOT
0,M1 TO AMOUNT(M1),M1
490  L = L + 1
500  B(L) = I
510  GOTO 340
520  IF J - M < 2 THEN 550
530  IF AMOUNT(M) < AMOUNT(M + 1) THEN 550
540  HCOLOR = 0 : HPLOT 0,M TO 100,M : HPLOT
0,(M + 1) TO 100,(M + 1) : AMOUNT(0) =
```

```
AMOUNT(M) : AMOUNT(M) = AMOUNT(M + 1) :
AMOUNT(M + 1) = AMOUNT(O) : HCOLOR = 3 : HPLOT
0,M TO AMOUNT(M),M : HPLOT 0,(M + 1) TO
AMOUNT(M + 1),(M + 1)
550  M = B(L) + 1
560  L = L - 1
570  IF L > 0 THEN 340
600  PRINT "DONE"
700  REM  QUICK SORT IS THE FASTEST I'VE SEEN,
BEATING OUT ALL EXCEPT FOR INSERTION SORT IN
UNDER 20 ITEM RANGE.  TIME FOR QUICK SORT OF
130 ITEMS - 35 SECONDS.  SEE OVERHEAD NOTE IN
LINE 305
710  REM  ORIGINAL ALGORITHM BY C.A.R. HOARE,
1962  (AN EXCHANGE SORT)
```

As you run the program, you will see that it picks a line halfway through the bunch and moves everything larger than this pivot point below it and the smaller lines above it. Then it goes back up to the one-quarter way point and does the same sort of moving. It keeps taking half the interval and moving the lines until it's working on individual lines.

You can actually use these two methods if you want a sorting program. If you want to alphabetize words instead of putting numbers in order, just change all the AMOUNT( ) to AMOUNT$( ) and, in the Insertion Sort, also change TEMP to TEMP$. You also can remove the lines above 900 and remove all the graphics statements that use HCOLOR and HPLOT. You will then have a sorting subroutine to use in any other program.

The Insertion Sort is handy for keeping records on things that you add to frequently, sorting the list as you go.

SORTING MORE THAN ONE THING

Suppose you have to sort something, but there is some information that goes along with it. For a simple example, let's assume that you have a list of people and their scores on an ESP mind-reading test that you have given to see how well they were able to guess some random numbers. You would like to alphabetize this list and print it out with their scores. Later, you might like to sort the list of scores and find out who got the highest.

In the sorting demonstrations above, you saw how long it took to move lines around. Information takes even longer. One way to sort is to not move the information, but just change the numbers on a list that tells you what order things come in. This is called **indexed array sorting**. Let's work first on the Insertion Sort, since that's easier to understand. Here's a demonstration program:

```
1  REM  INDEXED INSERTION SORT DEMO, SORTING
NAMES AND SCORES
5  NUMBER = 7
10  DIM INDEX(NUMBER + 1),AMOUNT$(NUMBER +
1),RESULTS(NUMBER + 1)
20  FOR ENTRY = 1 TO NUMBER : INDEX(ENTRY) =
ENTRY : READ AMOUNT$(ENTRY) : READ
RESULTS(ENTRY) : NEXT ENTRY
30  FOR ENTRY = 1 TO NUMBER : PRINT
AMOUNT$(INDEX(ENTRY));" -
";RESULTS(INDEX(ENTRY)) : NEXT ENTRY
40  GOSUB 900
45  PRINT
50  FOR ENTRY = 1 TO NUMBER : PRINT
AMOUNT$(INDEX(ENTRY));" -
```

```
";RESULTS(INDEX(ENTRY)) : NEXT ENTRY
60  END
100  DATA "JOHNSON, MARY",78,"HARRIS,
MARK",73,"PAULSON, MARION",113,"AARONSON,
JENNIFER",100,"ZEITLIN, FLORENCE",98,"YEAGER
JIM",69,"BILLINGS, RICHARD",89
900  REM  INDEXED INSERTION SORT(TAKES IN ARRAY
'AMOUNT$( )', STORES ORDER OF ENTRIES IN ARRAY
'INDEX( )'
910  FOR TIMESTHROUGH = 2 TO NUMBER
920  LET TEMP$ = AMOUNT$(INDEX(TIMESTHROUGH)) :
ITEMP = INDEX(TIMESTHROUGH)
930  FOR PLACE = TIMESTHROUGH - 1 TO 0 STEP - 1
940  IF AMOUNT$(INDEX(PLACE)) < TEMP$ THEN 970
950  INDEX(PLACE + 1) = INDEX(PLACE)
960  NEXT PLACE
970  INDEX(PLACE + 1) = ITEMP
980  NEXT TIMETHROUGH
990  RETURN
```

As you can see, in line 20, I set up an index in an array of where things are to begin with and then print them out to show you the original order. In line 920, I need a new temporary variable, ITEMP, to hold temporary values for my index. Notice that lines 950 and 970 do not move the AMOUNT$ ( ) around but just change the numbers in the index instead. Then I print the list out again, in its alphabetized order.

You could change this to sort the numbers first and add the names pretty easily, just by taking off the dollar signs in the sort subroutine and sticking RESULTS in wherever AMOUNT$ is and the reverse.

Of course, if you want the fastest method of sorting, here is the same program using the Quick Sort algorithm. Try just modifying the Quick Sort lines you already typed in and saved before, since only 1010, 1040, 1050, 1090, 1140, and 1150 need changing from your line-sorting skeleton.

```
1  REM  INDEXED QUICK SORT DEMO
5  NUMBER = 7
10  DIM INDEX(NUMBER + 1),AMOUNT$(NUMBER +
1),RESULTS(NUMBER + 1)
20  FOR ENTRY = 1 TO NUMBER : INDEX(ENTRY) =
ENTRY : READ AMOUNT$(ENTRY) : READ
RESULTS(ENTRY) : NEXT ENTRY
30  FOR ENTRY = 1 TO NUMBER : PRINT
AMOUNT$(INDEX(ENTRY));" -
";RESULTS(INDEX(ENTRY)) : NEXT ENTRY
40  GOSUB 900
45  PRINT
50  FOR ENTRY = 1 TO NUMBER : PRINT
AMOUNT$(INDEX(ENTRY));" -
";RESULTS(INDEX(ENTRY)) : NEXT ENTRY
60  END
100  DATA "JOHNSON, MARY",78,"HARRIS,
MARK",73,"PAULSON, MARION",113,"AARONSON,
JENNIFER",100,"ZEITLIN,
FLORENCE",98,"YEAGER,JIM",69,"BILLINGS,
RICHARD",89 JIM",69,"BILLINGS, RICHARD",89
900  REM  QUICK SORT
910  DIM B(22) : REM  ACTUALLY NEEDS 2LOGN
(BASE2).  22 IS GOOD ENOUGH FOR 2000 ITEMS.
920  L = 1
930  B(L) = NUMBER + 1
940  M = 1
950  J = B(L)
960  I = M - 1
970  IF J - M < 3 THEN 1130
980  PIVOT = INT ((I + J) / 2)
990  I = I + 1
1000  IF I = J THEN 1070
1010  IF AMOUNT$(INDEX(I)) < =
AMOUNT$(INDEX(PIVOT)) THEN 990
1020  J = J - 1
1030  IF I = J THEN 1070
1040  IF AMOUNT$(INDEX(J)) > =
AMOUNT$(INDEX(PIVOT)) THEN 1020
1050  INDEX(0) = INDEX(I) : INDEX(I) = INDEX(J)
: INDEX(J) = INDEX(0)
```

```
1060  GOTO 990
1070  IF I >  = PIVOT THEN I = I - 1
1080  IF J = PIVOT THEN 1100
1090  INDEX(0) = INDEX(I) : INDEX(I) =
INDEX(PIVOT) : INDEX(PIVOT) = INDEX(0)
1100  L = L + 1
1110  B(L) = I
1120  GOTO 950
1130  IF J - M < 2 THEN 1160
1140  IF AMOUNT$(INDEX(M)) < AMOUNT$(INDEX(M +
1)) THEN 1160
1150  INDEX(0) = INDEX(M) : INDEX(M) = INDEX(M
+ 1) : INDEX(M + 1) = INDEX(0)
1160  M = B(L) + 1
1170  L = L - 1
1180  IF L > 0 THEN 950
1190  RETURN
1200  REM  ORIGINAL ALGORITHM BY C.A.R. HOARE,
1962 (AN EXCHANGE SORT)
```

You won't notice much speed improvement on this short example. In fact, the Insertion Sort is actually faster for lists less than twenty items long. However, once you get above that, theQuick Sort is indeed quicker.

A CAUTION ON DATA ENTRY

One final note: if you have to sort something like a date or a serial number, or anything with nonnumerical characters in them, use the string sorting method [the one with AMOUNT$( )]. For example, if you had the five dates

9/26
8/13

9/4
9/14
12/25

A numerical sorter would ignore everything after the slash (/). On the other hand, the computer just sorts things from left to right. If you sorted this using the string sort version, the result would be this:

12/25
8/13
9/14
9/26
9/4

What happened? The computer looked at the first thing in each line. 1 is less than 8, even though we know that 12 is greater than 8. The same thing happened with those September dates. The 1 of the 14 is less than 4. If you are going to sort these as strings, you have to make sure they are all the same length. Thus, you should write them like this:

09/26
08/13
09/04
09/14
12/25

Then the computer would sort it as

08/13
09/04
09/14
09/26
12/25

Sorting is a powerful tool to use in science projects. It can help you recognize patterns and make sense out of masses of data. It can also make your lab reports and science projects much neater and more understandable to everyone else by having your data and results organized for easy interpretation. Effective communication of scientific results is one of the major goals of current science.

DATA BASES

In using an indexed sort such as the program above, you have just worked your first **data base.** A data base is a program that keeps track of things for you, just like a card file. The difference is what you can do with the information on the cards later.

For example, let's use the project I started this chapter with—the heights of trees in your city. In your notebook, or right on a card, for each tree you could put down the type of tree, how tall it was, where you found it, how old it was (sometimes you can tell that by how many branch groups there are, such as on fir trees), whether it was an evergreen or a deciduous tree (one that loses its leaves each year), and perhaps when you wrote down the information.

When you got home, you could transfer the information to a data base program. When you start up one of these, you have to define the "fields" or categories you wish to use to hold your information. Usually, you have to tell the maximum amount of information you want to put in each category. For example, you could specify NAME as one field with a length of thirty letters maximum. Then you would give similar information for HEIGHT, LOCATION, AGE, EVERGREEN/DECID., and DATE.

Once you had typed in the names for each field, you could then enter your tree data for each tree. So far you haven't done anything different from a card file. If you filled out each card as you looked at the trees, your cards would probably be in date order. What if you wanted to alphabetize the cards by tree name instead? With the card file, you would have quite a job ahead of you. With a data base, you could tell the computer to "sort" on the NAME field. In a few seconds, it would be done. You could then print out all the trees with whatever part of the rest of the information you wanted. One card might look like Figure 7.

```
NAME: Acer Macrophylla    COMMON: Big Leaf Maple
                          TYPE: Deciduous
HEIGHT: 26 ft             LOCATION: G8
AGE: 10 yrs (est.)
```

Figure 7. A card file is a data base.

Instead of by name, you could tell the computer to sort by the HEIGHT field instead. In a short time you could then put out a list of the trees, with the tallest first.

Besides sorting, data bases let you specify how much gets printed out. For example, you could tell the computer to print out only the locations and ages of trees where NAME=MAPLE. You would then have an interesting study of what kinds of trees grew in which parts of the city. You could amplify it by saying that NAME had to equal MAPLE, the AGE had to be greater than 30 years, and the LOCATION had to be in map references G5 through G8. Then you would be able to tell if there were patterns in parts of the city where maple trees had all been planted around the same time.

This type of searching is called "ANDing," since you specify that all the conditions must be met: condition 1 AND condition 2 AND condition 3, etc. Besides the AND function, you could also OR conditions. You could tell the data base to print out the names of the trees where the AGE was greater than 20 years OR the HEIGHT was greater than 25 feet (or 10 meters or whatever units you used for all of the trees).

You or a programmer friend could use the indexed- sort ideas above to write your own data base. However, you could probably find one in old issues of some magazines. *SoftSide* and *Creative Computing*, to name just two, at one time published a data base by Mark Pelczarski of Penguin Software (called *Developing Data Base*) that was in Apple, Atari, and PET versions.

Commercial data bases are available for a wide variety of prices. You need to look at several in a store to see which will fit your needs. You need to check for how fast they sort, how many records you can have, how big each record can be, how many fields you can specify, how much ANDing and ORing you can do in the output, and how specifically you can tell the computer to make the printout look. If your school has some newer Apples, you may find that it already has a copy of *AppleWorks*, a program that combines a data base with a spreadsheet and a word processor but is fairly expensive.

Also for Apple is the capable but rather intimidating and expensive *dBase II*, and the friendlier but still somewhat expensive *pfs:file* and *pfs:report*, *List Handler*, and the moderately priced *Bank Street Filer*. There are numerous public domain data bases and many modifications to the venerable *File Cabinet*, which came on the Apple Master Disk with the machine.

IBM machines have a very large number of data bases, mostly oriented toward business, with the cheaper ones still fairly expensive. Look for names such as *dBase II*, *dBase III*, *Friday!*, and *pfs:file*.

Commodore machines have a more limited selection that I could find: the standard *pfs:file*, *Superbase 64* for moderate cost, and *File Now-64* which is somewhat cheaper.

Atari users can get *Synfile File Manager*, *FileWriter* and *Small Business Inventory* which can be adapted to a science file, both from Atari itself, and lastly *Mastertype Filer*. All of these are very inexpensive.

Data bases and sorting allow you to collect information, organize it many different ways, look for similarities and differences, and then test your conclusions on all the data in your file—observation, classification, hypothesis, and experiment: science on a microcomputer.

IDEAS FOR SCIENCE PROJECTS

A few companies are putting out collections of data to be used in science classes with some of the more widely known data base programs. Scholastic is the best known with its series of files in physical science, life science (and U.S. history and U.S. government too), for the *pfs:file* data base. These data base collections contain a lot of information about fish, mammals, plants, or something of that sort. Then all you have to do is play with the program to see what you can learn about things that seem to go together. Do the largest sea animals all eat the smallest food? What about the whale shark, blue whale, and right whale? Check with your science teacher for sources of these programs.

However, you can make up your own data base of information you collect. Use your tree collection to see significant patterns in growth of your city, a rock collection with a field which tells where you found a rock to see if there are patterns of mineral deposits, or notes on chemical reactions to see if any pattern of reactions with certain ingredients becomes apparent when you can print out things selectively.

The neat thing about data bases is that you can reprocess information to find out new things.

One project you might do is to enter into a data base information available from standard chemistry books. Put in the names of elements, their symbols, density, electrical conductivity, valence charge, melting point, boiling point, and ionization energy. Silver, gold, and copper all conduct heat very well. They also conduct electricity well.

Does this relationship hold for other metals? Sort the data by highest conductivity of electricity and then print out the heat conductivity of elements and see if these numbers go together. Test how well these data fit together using correlation analysis or another statistical method. Graph your results. Check to see if density has any affect. See which elements will float in a pool of mercury or molten tin.

Another project is to make a data base of inventions. Include the inventions, their inventors, the date of invention, the inventions that couldn't be realized until a later invention was made, the immediate impacts were, etc. Prepare a chronological list of inventions of a particular type, such as home/leisure, communication, or transportation. Include in your list the impact data. Then interview some older people (the older, the better) and ask them for their opinion as to how these inventions changed their lives or the lives of others. There are people alive today who have traveled as children in horse-drawn vehicles and flown in jets. Summarize this information by selecting three inventions that had the greatest impact on people or society. Prepare a chart that shows how the impact of these inventions spread to other fields?

What about the nationality of inventors? Can your data base provide numbers of inventions by nationality of inventor? Has the ratio of, for instance, American inventors to Japanese inventors, changed in the last 80 years? Have wars had an effect on this ratio? Have wars had an effect on the *types* of inventions? If you made a graph of significant inventions using a graphing program, would there be "clustering" of inventions about particular time periods?

5

SIMULATIONS

Simulations are what you do on a computer when doing the real thing is too much trouble, takes too long, happens too fast, is too dangerous, too expensive, or just plain impossible, or when the real thing is too huge to work with or too small to see.

SIMULATIONS ON SMALL COMPUTERS

One commercially available program allows you to simulate the orbits of the planets, from any distance away you choose, and from any angle. The calculations to show the view are not particularly difficult, but they are tedious and the computer can do them much more quickly than you can, and it can draw the results.

Another program allows you to design a sun with a planet. You get to specify the type of star and then some details about the planet—how big, how much mass, and how reflective the atmosphere will be. The program does some quick calculations and tells you how long the "year" would be and the temperature of

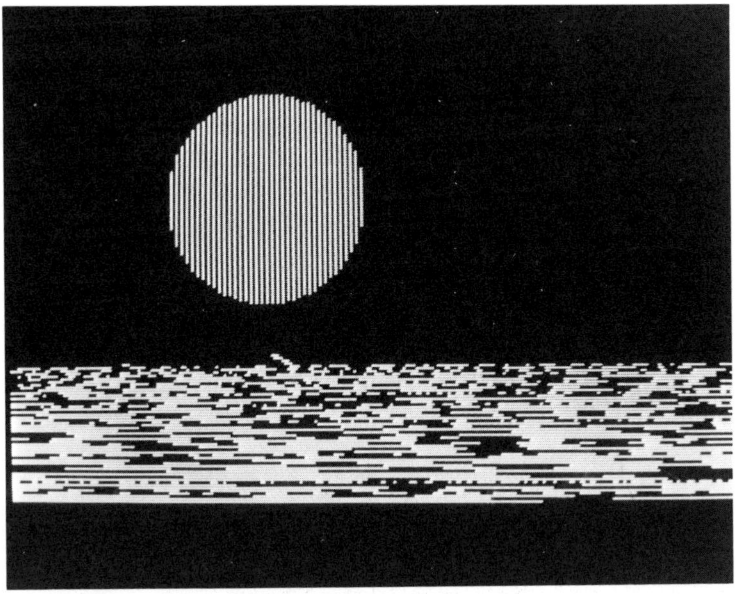

This simulation of an alien landscape was generated by Sheridan Simon's Astronomy Disk *from Prentice-Hall.*

the surface. Then a typical view from the planet is drawn on the screen. (Often, all you get is a charred plain with a huge sun baking a few mountains. You may have to move the planet much further away from the star, or try raising the reflectivity of the planet's atmosphere to turn away more of the heat before it can strike the surface.)

Making your own planet to try and achieve terrestrial conditions with a much different sun is difficult to do without a simulation! This is not a School Project or even a Saturday Project. Not even Carl Sagan or Mr. Wizard builds planets on the weekend any more.

A good simulation should be able to do what the real-life object would do, but in such a way that you can understand what is going on. If you understand it, then you should be able to make a prediction about what the real object would do in a given situation.

In Chapter 3, one of the first programs there simulates a series of yes/no questions by having a ball drop down through a triangle of obstructions, choosing to go left or right at each point. By using this simulation, you can see how, even with seven decisions, you still end up with results that approximte a bell-shaped curve (this is explained more fully in that chapter).

In *Rocky's Boots* (published by the Learning Company for a moderate price, for Apple, Atari, and Commodore), you are given a series of digital logic-circuit

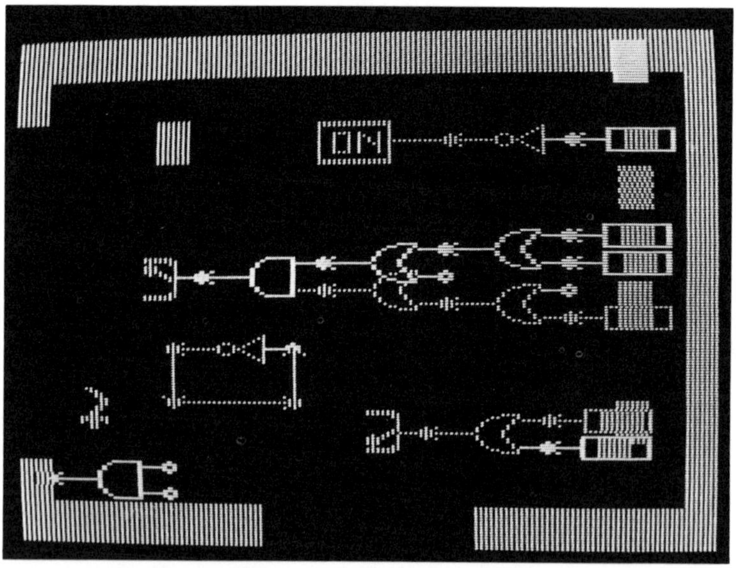

A screen from the Learning Company's Rocky's Boots *showing manipulation of AND gates, OR gates, and NOT gates. In addition to being fun, this simulation can teach you about logic and electronics.*

components. Logic circuits are named by their function, as they act as "gates" to allow electricity to flow or not. For example, a gate that requires two wires being turned on at once to make something happen is called an AND gate. One where either wire works is an OR, and so on.

You can buy these gates from electronics suppliers and hook them up and find out what happens when you put a whole bunch of them together. *Rocky's Boots* lets you move pictures of them around on the screen to connect them, and the program makes the symbols act the same as the real thing would.

The program has excellent graphics and is fun to run. You also learn a lot from it about digital logic circuits. A later program from the same company, *Robot Odyssey*, is much more complex, and you start designing whole electronic circuit chips, using a lot more complicated gates.

This is exactly what electronics firms around the world today do in their research departments.

Nobody lays out the chips and gates on the table first and hooks them up. Using sophisticated computer programs, the electronics engineers move components around on the screen, using a light pen, tablet, or joystick, and simulate the operation of the finished circuit before they ever invest the time, money, and hardware in actual construction.

Cat Lab (somewhat expensively priced from Conduit) simulates the breeding of several generations of cats. If you cross a purebred white cat with a purebred black cat, what will the next generation of kittens look like? What if you keep on breeding litters with more and more variables, such as long hair or short, with blue eyes or yellow, with a long tail or short, etc.? I'm not saying that breeding cats is not a good way to learn about genetics, but most science fair programs will not allow you to use animals more complicated than insects. Besides, perhaps an animal project is more ambitious than the local SPCA, your time, or your capacity for emptying litter boxes allows. However, with the computer, you learn the same things about genetics without the mess and time involved otherwise, but still *doing* experiments, rather than just reading about them.

Although all of the programs above use graphics heavily to make their concepts clear, at the heart of them they are all mathematics, describing how real-world things interact with their environment. In many cases in science, we are able to describe how something works, such as how a glacier flows, a river changes, or how planets move in their orbits, but we can't afford the time it would take to get some real information from the system. By making the computer speed up time, we can learn about many things and predict future events.

For example, take the movements of the solar system. Not all asteroids are in that belt between Mars and Jupiter. Some come very close to the Earth. *Project Icarus* is named after one of them. By taking into account the gravitational pull of the sun, and all the planets and other asteroids, computer programs are right now trying to simulate the motion of these closely approaching massive bodies to see if one will hit Earth. If one of them did, the resultant dust from explosions could block enough sunlight to cause another ice age, similar to the predictions of a "nuclear winter." There are too many factors to make a prediction without the fast power of the computer to speed up time, to give us enough warning to see if we can do something to avoid such a collision.

DESIGNING A SIMULATION

Simulating a scientific event requires careful observation and clever thinking to come up with the mathe-

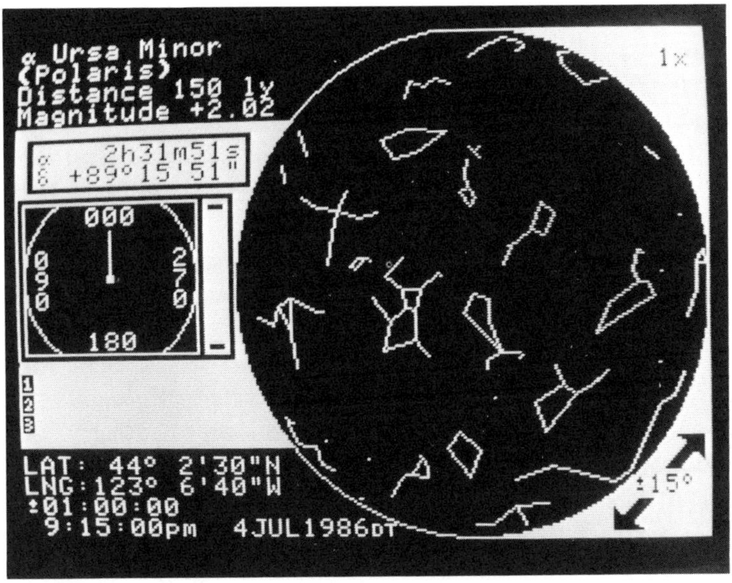

Lightspeed Software's The Observatory *shows the night sky for July 4, 1986, at 9:15 p.m. above Eugene, Oregon.*

matics needed to mimic something's behavior. Luckily, there are "curve-fitting" programs around to help. In these, you put in data from an experiment and the program graphs them. Then it draws a smooth line or curve that goes through as many of the points as it can.

There are several ways to do this. You can try a straight line, which would have the mathematical form $\mathbf{Y} = \mathbf{mX} + \mathbf{b}$, just like you've seen in your beginning algebra class, where **m** is the slope of the line and **b** is where it intercepts the **Y** axis (The "slope" is how steep the angle of the line is. A slope of 2 would go up 2 units for every 1 unit to the right. An **m** equal to 5 would be steeper, going up 5 for every 1 over).

The program should be able to give you the **m** and **b** for any line it draws. If that line doesn't really seem to fit the data, you might try a nonlinear (curved) line, such as a quadratic of the form $\mathbf{Y} = \mathbf{aX^2} + \mathbf{bX} + \mathbf{c}$, with the computer drawing the line and giving you the coefficients **a**, **b**, and **c**.

A very inexpensive commercial program that does this type of function well is *Graphical Analysis II* by Vernier Software.

The curve that shows up on the screen from your data might be exponential, logarithmic, or something else you may not have ever heard of yet in your algebra class. Once you get an equation, you can write the program and the graphics to do some predictive work. Learning what various mathematical functions look like when graphed can speed up your selection greatly.

If you want to have a lot of fun and get a real feel for how equations work, try running *Algebra Arcade* (Wadsworth Electronic Publishing), moderately priced for Apple, IBM, and Commodore computers. You get a playing field with little alien "algebroids" on it and have to shoot them down. You hit them by typing in an equation at the bottom and seeing what happens. In one or two sessions of exciting play, you can learn a tremendous amount of algebra and trigonometry that will help you in math and science courses for the rest of your career.

How can you use simulations on a computer in a science project? One way might be to write your own program. To do that, you will have to know how something operates quite well, in order to make up an algorithm for how it operates.

Real life is not always predictable. One can predict some broad general things, but the day-to-day stuff is just too hard and appears to happen "randomly." All versions of the BASIC language on computers of all sizes have some way to generate random numbers. To make many simulations, it is necessary for you to

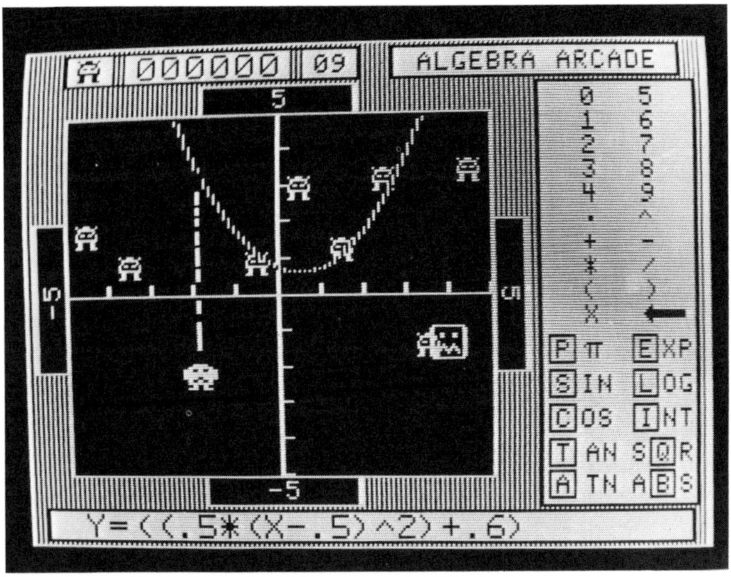

Algebra Arcade, *from Wadsworth Electronic Publishing, is a mathematical game that teaches the shapes of curves. In this setup, the equation* $Y=0.5(X-0.5)^2+0.6$ *is being used to shoot down four alien "algebroids."*

throw in random factors, to see how things would operate in the real world.

If you take measurements of how many cars go by a certain place in your town (or got the information from the city street department), you might be able to write a program that would simulate traffic flow at different times of the day. You might use a method that kept adding numbers to street intersections, represented by a variable or array, somewhat the way I kept increasing the "slots" in the bell-shaped-curve simulator. A busy main street might add cars at three times the rate of a side street. What happens if you block one lane for road work? You could shift numbers from one street to the others nearby, keeping the total number of cars the same. However, you will also have to include a random factor for the traffic flow. Traffic may have an average rate at certain times of the day, but the random part could be for moment to moment.

MORE THAN JUST MATH —RANDOM NUMBERS

Genetics programs that predict how plants will reproduce are based on Mendel's laws of genetics. But individual plants sometimes vary within those laws. Mendelian genetics may predict that the second generation of crossbred tall and short pea plants should be 25% pure short, 25% pure tall, and 50% crossbreeds, but in reality, each time you do this, you may come up with slightly different numbers. Using a random factor more closely simulates real life.

IDEAS FOR SCIENCE PROJECTS

It's difficult, at this level of science, to come up with projects you can do. A well-done simulation can be of great value in demonstrating one of your projects to someone else, but you have to do a project first. A carefully constructed simulation by someone else can be used to learn things you didn't know about a process, but it won't truly be new material for the rest of science.

Scientists use mathematical simulations to check to see if they understand a process. For example, by observing stars carefully and seeing the color of stars they know the temperature of, they may be able to make a mathematical model of the nuclear and photonic reactions involved so that they can relate color to temperature. Then they will run their simulation to get a result for a new temperature and then check it against a star they know both temperature and color on. If the results agree with the simulation, they then know they understand what is happening.

Some ideas can be found in earlier parts of this chapter that may give you some ideas, but making good simulations at your level of scientific knowledge and computer expertise can be difficult.

6
THE COMPUTER AS A LABORATORY TOOL

The computer has a great use in the laboratory in collecting data. It is the tireless assistant, the meter reader who's always there on time, the lab assistant who is accurate, fast, and sees everything.

There is a fairly easy experiment you can do at home, using a candle, a test tube, a thermometer, and some mothball flakes. All you do is heat up the mothball flakes (chemical name: paradichlorobenzene) in the test tube over the candle until they are all melted, put the thermometer inside, and then write down the temperature every 30 seconds.

The objective of this experiment is to find out the freezing point (which is also the melting point) of paradichlorobenzene. You might think you could tell that just by watching closely. But, consider this—as you watch the cooling take place, do you count the freezing point as when the very first crystal starts to appear, when the very last liquid solidifies, or someplace in between? If it's in between, *where* in between?

If you were to plot the temperatures you record on graph paper (or use the program from Chapter 2), you would get something like Figure 8.

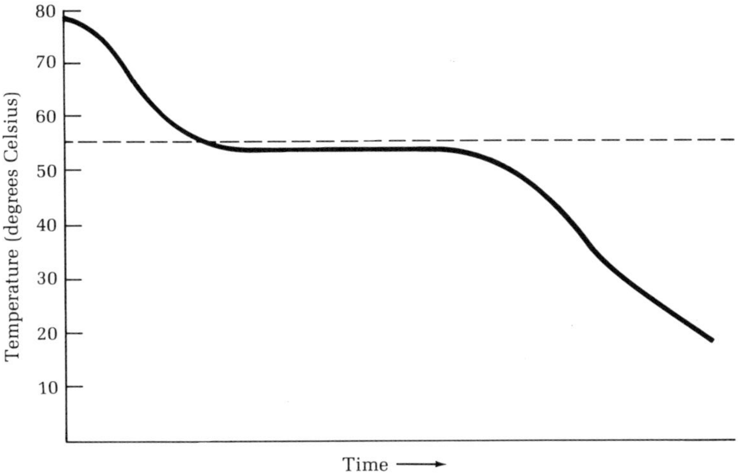

Figure 8. Cooling curve for mothball flakes (paradichlorobenzene)

The melting/freezing point is the flat part of the curve. This is easy to see on a graph but not always obvious when you are writing down the temperatures.

THE COMPUTER AS A DATA RECORDER

You could use a computer as your data recorder. Instead of writing down the temperatures and then typing them into a graphing program, you could let the computer measure the temperatures, store those for you for later printout, *and* graph the temperatures all at the same time. The computer will be more accurate, won't make mistakes recording the numbers and can do it quickly enough to take temperatures faster than you could write, such as once every second.

Scientists have adopted this use of the computer with great enthusiasm. They are not all using highly sophisticated computers either. Many labs are using microcomputers just like you have in school or at home.

I visited a chemical testing lab this year where they make adhesives for making plywood and particle board. One of the tests they do is to glue small blocks of wood together with their newest adhesive, use a machine to tear the two pieces apart, and then examine which failed first—the glue or the wood. By using different amounts of adhesive and looking at how much wood is still left on one half, they can find out how strong the glue is and how little they can use and still make it stronger than the wood itself.

In the past, a person would sit at a desk, writing down "percentage wood" columns of numbers, keeping track of which sample was being examined, then later do some calculations on each set of numbers which would relate thickness and strength, and then later cross-check these with the original specifications on each block. Now the work station has a computer at it.

The thickness information and block details are typed into the computer *as they are made*. The computer does the calculations immediately, typing out all relevant information on a printer attached to the computer. Some of the data is stored on disk and later graphed.

Eliminating all the steps of writing down the numbers, recopying them later, and then using a calculator for the tests has increased accuracy and speed quite a bit. But the computer can go one step further. The computer can *collect* the data in the first place.

COMPUTERS AS MEASURING DEVICES

Because early pioneers in microcomputers wanted to play games, just about every machine has a place to

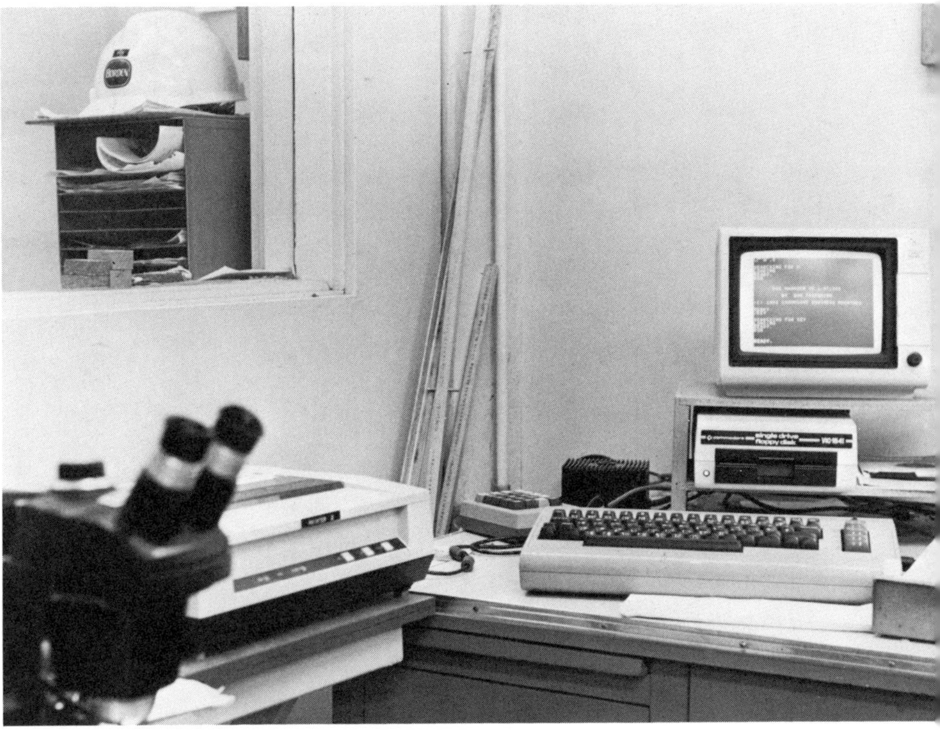

Commodore 64 being used for data entry in an industrial lab.

plug in a joystick or paddles. On the Apple II+ and IIe, the place where you put the sixteen-pin paddle plug is inside the machine.

Instead of opening the lid of the machine up each time you insert the plug, you can get a cable to move the plug to the side of the machine, such as the way the machine in the top photograph on page 94 has been modified. Most computers, such as the Atari and Commodore, now have moved the joystick port to the outside of the machine, as have the Apple IIe and IIc. Check the manual for your machine to find the various input ports.

E Z PORT
VERSA COMPUTING INC.

Although more convenient to use, you may be limited to only two paddles, or one joystick, or, in our case, only two scientific probes.

The paddle is a device which is able to vary electrical resistance. A special circuit in the computer then takes this resistance and changes it to a number. Such a function is called an *analog-to-digital input*. If you have paddles or a joystick, type in this program:

```
10 REM PADDLE TESTING PROGRAM
20 HOME
30 PRINT PDL (0), PDL(1)
40 GOTO 20
50 END
```

For an Atari, you just need to make the HOME command PRINT CHR$(127), and use PADDLE(0), PADDLE(1) instead of PDL(0), PDL(1). Other machines may use STICK(0) or a particular PEEK.

As you run this, you will see that a number between 0 and 255 appears on the screen for each paddle. If you are using a joystick, the left column will correspond to the X-axis and the right column to the Y-axis. Inside the paddle or joystick are resistors arranged on a wheel. As you move the wheel, electricity has to go through more and more resistors. The computer reads how much of the electricity it expects

Above: Apple II+ with 16-pin game ports. The Apple IIe also has these ports. Below: Back of Apple IIc showing 9-pin game port. The Apple IIe also has this port.

is getting through and assigns it a value on a scale from 0 to 255.

Other computers, such as the Atari, may use a different system, such as four digital switches for a joystick that increase a counter in the computer each time the stick is pressed in one direction. The effect is similar, however.

RECORDING THERMOMETERS

Electricity travels in a wire when electrons jump from one metal atom to another. Electrical resistance is an indication of how difficult this transition is. This ease of jumping is sometimes dependent on how hot the metal is. A device called a **thermistor** is made of wires of two different metals joined together and changes its resistance as the temperature changes. By putting a known quantity of electricity through this wire (usually 5 volts) and seeing how much comes back, the computer can again change a continuously variable reading (analog) into a number (digital). If you calibrate this by using two temperatures, say, freezing water and boiling water, you can figure out a ratio between 0 to 100 degrees Celsius and 0 to 255 on the paddle values. Instead of your reading a thermometer and typing in numbers, your computer could do both.

Several fine commercial programs do this quite well. For the Apple, there is *Microcomputer Physics Lab—Heat*, (somewhat expensive from Cross Educational Software). This kit lets you make three temperature probes and has programs to read all three probes at the same time, printing out the temperatures and graphing all three temperature readings as they are read. You can see what's happening *as* it happens.

HRM Software sells a similarly priced *Temperature Grapher*. You get only two probes, and the program will only read two at a time, but these come already assembled and just need to be plugged in.

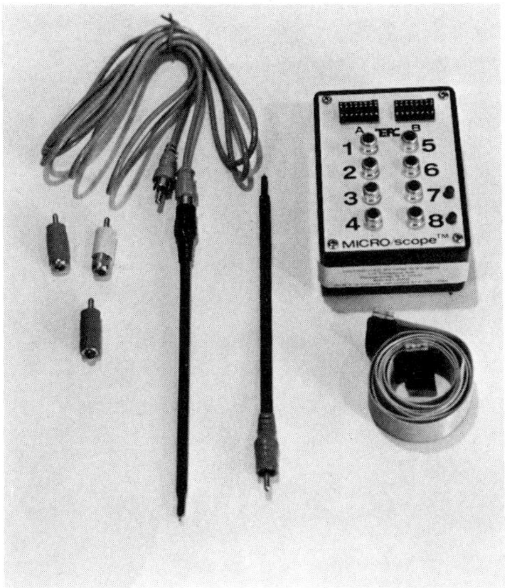

Interface module and sensors from HRM Software's Experiments in Science.

The temperature probes are part of two larger packages. You can get them in HRM's rather expensive *Experiments in Human Physiology* or *Experiments in Science.* The sets also contain light probes, small light sources, and a junction box, containing an extra paddle receptacle on top so that you can still have your joystick plugged in when you're not using the scientific apparatus. Even though these are too expensive for home use, there's a chance your science department might already have them or would be able to order them for use by several classes.

Vernier Software has reasonably priced software that you can modify as you wish. *Temperature Plotter* is one such setup.

You *could* try to find thermistors of the right resistance for your machine (0 to 150 kilo-ohms on the Apple), figure out where the probes connect, and hope you don't short-circuit a $1,000 computer if you have the wires hooked wrong. If you know how to solder

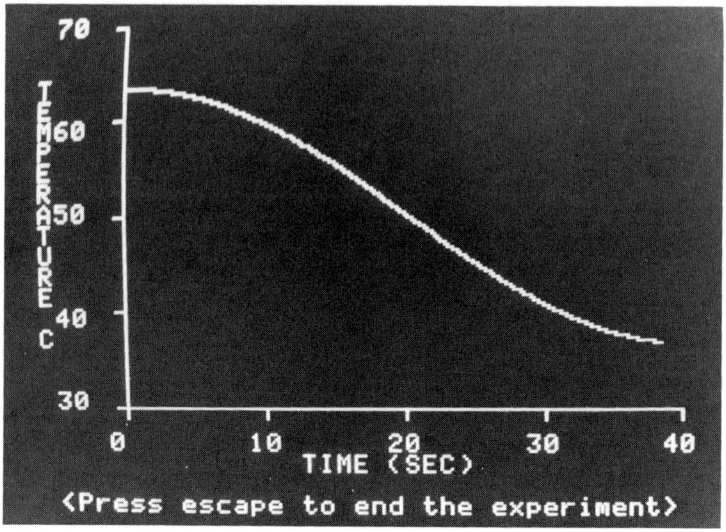

This cooling curve was generated by HRM Software's Temperature Grapher *while temperature probes were connected to the computer.*

two wires together, the kit from Cross would give you the most capability.

However, in writing your own program you might not get the timing loops done properly. They can get somewhat tricky. Accurate timing is very important for this type of experiment. In the commercial versions, you can set the intervals between readings of the thermometer to be a range of numbers. For example, in the Cross package, the graph will hold 270 points. You can set the time interval between readings to a number of seconds between 1 and 1,000. That means your graph could display 4.5 minutes (270 seconds) to *75 hours* (270,000 seconds or over 3 days)!

For the Atari, *Atarilab Starter Set* (developed by Dickenson College, Atari Corp.) has one probe for moderate cost. The temperatures you can use with it, however, are rather limited. The accompanying book says that the hottest water you should use is "as hot as bath water." Software and hardware packages such as this might severely restrict the types of experiments you could do.

USING DIFFERENT DEVICES

What *can* you do with temperature probes and a computer? One thing might be to test various insulating materials you devise. You could put a cup of hot water with one probe inside of your material and another outside to record the temperature difference over a period of a few minutes, a few hours, or a few days, depending on how good the insulation is. Perhaps you might wish to use three probes at various rooms in your house to see what the "microclimates" were as the day progressed. You might even find variations within one room, depending on how the light from an open window on a sunny day affected things.

Many microcomputers can send signals *out* of the computer to turn things on or off, usually using a fairly standard interface. The "BSR" interface is one sold in hardware stores for controlling lights and small appliances from one location, making it perfect for your computer.

Some kits, such as HRM's *Experiments in Science*, include a "light probe." This utilizes a substance, such as cadmium sulfide, that changes its resistance as light falls on it. You could use a "photoresistor" that you can buy from electronics stores to tell when the light gets to a certain level and sends a signal out to turn on a light. Or you could use a temperature sensor to see when a house gets too hot during a summer day and send a signal to an electric motor to unroll an awning over the windows on one side of the house. A scientist might use a temperature sensor to see how a reaction is going and then have the computer turn on a stirrer, cooling bath, or valve to add more chemicals.

These interfaces usually come from "slots" in the back of the computer. Other machines may use different approaches. For example, the Atari has four ports for joysticks on the outside that some companies have used to make interfaces similar to the Apple's eight slots, IBM's four slots, and Commodore's one.

In a chemical lab I visited, the setup I saw in the photograph on page 101 used an Apple. The large apparatus is called a gas layer chromatograph. If you boil a liquid that has many components and blow the vapors down a tube with helium gas, the components will separate out over the distance. Those components with the lower boiling temperatures will arrive at the end of the tube first; the higher-boiling-temperature ones will take longer to reach the end. The machine senses when something goes past the electrodes and tells the computer. The program keeps track of how long it takes a given component to pass the electrodes and converts this to a percentage of the total. It stores the numerical data but immediately shows a graph on the screen. The numbers can be printed out and the screen "dumped" to the printer as needed. Figuring out what's in a new mixture now takes minutes instead of days, as it used to before computers appeared in the labs.

THE SMART STOPWATCH

Timing devices are useful to scientists. Physicists need precise measurements of time for falling objects, accelerations, velocities, and rotations. Chemists need

Interfacing this Apple computer with a gas layer chromatography apparatus allows the reading of concentrations of chemicals in mixtures. The analog-to-digital interface converts information on the mixture being analyzed into a form (digital) understandable by the computer.

to study how fast certain reactions can work and when to stop others. Biologists time the movements of plants, animals, and even the growth of bacteria. Using a stopwatch has its limitations, both in convenience for the particular task and accuracy.

The computer itself has a clock in it to coordinate each activity: shifting numbers around, making sure memory locations are emptied in time to get a new number, timing the output of characters to a television screen so that they correspond to the lines there, refreshing those images, etc.

To give you an idea of how fast the computer can count, try this sound program for an Apple:

```
10 REM SOUND PROGRAM
20 CLICK = −16336
30 FOR TIME = 1 TO 30
40 A = PEEK(CLICK)
50 NEXT TIME
100 END
```

Other machines can do similar things. For example, for the Atari, change line 20 to CLICK = 53279, and line 40 to POKE CLICK,0.

Memory location −16336 (53279 on the Atari) is a special place for the computer that makes a click on the speaker when you look at it (PEEK) or put a number in it (POKE). Line 30 starts a loop that PEEKs the speaker 30 times. What you hear when you run this program is a series of clicks, fast enough so that they are perceived as a tone.

You can modify this to control how long you wait in between each click and so change the tone. Here, add these lines:

```
15 INPUT HOWLONG
45 FOR DEELAY = 1 TO HOWLONG : NEXT
DEELAY
60 GOTO 15
```

Now when you run the modified program, typing in a small number will make the clicks come close together. Just as a guitar string vibrates faster for higher pitches, you should hear the clicks making a higher-frequency sound. Big numbers will slow it down—a lower pitch.

That's in BASIC, a relatively slow way to make the computer work. The computer needs to interpret BASIC into machine language (base-16 numbers like

F0 FF 30 20 that the computer actually can use). If I were to write the same program above in machine language (the computer's "native" language of 0's and 1's), it would run so fast that the clicks would become a tone you would recognize as a very high pitch. That's how computers make music. The point is, this extreme speed can be used to time things very accurately, and to record that data and do calculations with it.

THE COMPUTER AS TIMER

You can use this speed to make a quite accurate stopwatch. However, there should be some reason to use a computer instead of a watch. The program below will display several times on the screen and even give you a running average. You could use this to time how long it takes an object to fall from a distance and thus find the acceleration of gravity at your location; time the speeds before and after collision of two balls of glass, plastic, rubber, or ivory and figure out the coefficient of elasticity; time insects in a maze repeatedly to see how long it takes them in successive runs as they learn it; or even how long it takes another computer to do calculations using different methods.

```
5  REM  BASIC TIMER
10  D$ =  CHR$(4) : PRINT D$;"OPENTEMP" : PRINT
D$;"WRITETEMP" : POKE 33,33 : LIST : PRINT
D$;"CLOSE" : POKE 33,40 : END
20  KEYBOARD =  -16384 : RESETKBYBOARD = -16368
25  HOME : VTAB 12
30  PRINT "(C)alibrate, (T)ime, or (Q)uit"; :
GET CHOICE$ : PRINT CH$ : IF CH$ <  > "C" AND
CH$ <  > "T" AND CH$ <  > "Q" THEN 30
40  IF CHOICE$ = "C" THEN GOSUB 500
50  IF CHOICE$ = "T" THEN GOSUB 100
60  IF CHOICE$ = "Q" THEN END
```

```
70  GOTO 30
100  REM  TIMER ROUTINE
110  IF FACTR = 0 THEN  PRINT "Run the
(C)alibration first" : FOR PAUSE = 1 TO 2000 :
NEXT PAUSE : GOTO 270
120  HOME : VTAB 1 : PRINT "PRESS <SPACE BAR>
TO START OR STOP" : TRIAL = 0 : SUM = 0
130  TRIAL = TRIAL + 1
140  VTAB 2 : HTAB 1 : PRINT "START TIMER : ":
CALL -868 : REM  CLEAR TO END OF LINE
150  GET K$ : FLASH : VTAB 2 : HTAB 20 : PRINT
"TIMING" : NORMAL
160  COUNT = 0
170  COUNT = COUNT + 1
180  LOOK = PEEK (KEYBOARD) : IF LOOK > 127
THEN 200
190  GOTO 170
200  POKE RESETKEYBOARD,0
210  IF CHOICE$ = "C" THEN 270 : REM  TO THE
RETURN
220  TIME = COUNT * FACTR
230  SUM = SUM + TIME : AVERAGE = SUM / TRIAL
240  VTAB (TRIAL + 4) : HTAB 1 : CALL -868 :
PRINT "#";TRIAL;" ";TIME;" S.  AV.=";AVERAGE;"
S."
250  VTAB TRIAL + 5 : HTAB 1 : PRINT "AGAIN
(Y/N)"; : GET ANSWER$ : PRINT ANSWER$ : IF
ANSWER$ <  > "Y" AND ANSWER$ <  > "N" THEN 250
260  IF ANSWER$ = "Y" THEN 130
270  RETURN
500  REM  TIMER CALIBRATION
510  HOME : VTAB 20 : PRINT "Press <SPACE BAR>
to start.  At the end of 10 seconds, press
<SPACE BAR> again. We will do 3 trials and
average." : SUM = 0
520  FOR CTRIALS = 1 TO 3
530  GOSUB 140
540  SUM = SUM + COUNT
545  VTAB CTRIAL + 15 : HTAB 1 : PRINT "TRIAL
#";CTRIAL;" COMPLETED"
550  NEXT CTRIAL
```

```
560  CAVERAGE = SUM / CTRIALS : FACTR = (10 /
CAVERAGE) / 1.33333
580  RETURN
```

The program works fairly well, although you really have to calibrate it each time. If you add to the program in any way, the calibration may change, since more will have to be held in memory by the computer and it may not get to each line as quickly.

To really get accuracy, we need to rewrite the program with the actual counting being done in a machine language routine. Instead of having you go into the monitor and typing all those hexadecimal numbers, this version POKEs the machine language program in and CALLs it from the BASIC program:

```
5  REM  MACHINE LANGUAGE TIMER
10  D$ = CHR$(4) : PRINT D$;"OPENTEMP" : PRINT
D$;"WRITETEMP" : POKE 33,33 : LIST : PRINT
D$;"CLOSE" : POKE 33,40 : END
20  FOR LOC = 1 TO 62 : READ BYTE : POKE 767 +
LOC,BYTE : NEXT LOC
30  HOME : PRINT "HOW MANY RUNS DO YOU WISH TO
DO?" : INPUT HOWMANY
40  SUM = 0
50  HOME : INVERSE : PRINT "PUSH BUTTON TO
START, RELEASE BUTTON TO STOP." : NORMAL : POKE
34,5 : PRINT : PRINT
60  FOR EVENT = 1 TO HOWMANY
70  PRINT "TRIAL # ";EVENT;" : ";
80  Y = PEEK (49249) : IF Y < 128 THEN 80
90  CALL 768
100 TIME = (15 * PEEK (830) + 3847 * PEEK (831)
+ 985105 * PEEK (832) + 110777) / 1022714
110  SUM = SUM + TIME : AVERAGE = SUM / EVENT
120  PRINT TIME;" S."; : HTAB 26 : PRINT "AVG=
";AVERAGE
130  NEXT EVENT
```

```
140  POKE 34,0 : PRINT "ANOTHER RUN?"; : GET
ANSWER$ : PRINT ANSWER$ : IF AN$ <  > "Y" AND
ANSWER$ <  > "N" THEN 140
150  IF ANSWER$ = "Y" THEN PRINT "CAUTION -
SCREEN WILL CLEAR AND RUN WILL BEGIN WHEN YOU
PRESS A KEY."; : GET KEYPRESS$ : GOTO 30
800  REM  MACHINE LANGUAGE PROGRAM POKES
810  DATA
169,0,141,62,3,141,63,3,141,64,3,169,170
820  DATA
141,48,3,141,48,192,32,30,3,206,48,3,208,245
830  DATA
76,41,3,162,5,160,15,136,208,253,202,208,
248,96
840  DATA
173,97,192,16,15,238,62,3,208,246,238,63,3
850  DATA 208,241,238,64,3,208,236,96
```

To use this program, you have to have a paddle or joystick. Start the timing by pressing down the button on paddle #0. To complete the timing, just release the button. As the last program did, this one gives you a running average.

IDEAS FOR SCIENCE PROJECTS

With the program given above, you can add as many things as you want. For example, you might wish to INPUT the distance something has to move and get a printout of not only the time, but the speed (distance/time). Since the machine language part does all the counting and won't be affected by how long the program is, you can add as many things onto it as necessary.

An inexpensively priced example of a commercial program of this type for the Apple and Commodore 64 comes from Vernier Software. *Precision Timer* will time very accurately and can even put large numbers on the screen that can be seen from the back of the room if you are demonstrating timing to an audience. A kit is also available for adding up to three photogates to the timer. This will allow automatic timing by objects passing in front of the photogates. With three such gates, you can determine accelerations easily.

You could also use the photogates to tell where a laboratory animal goes in a cage or pen and when the event happened. You could do studies of animal activity this way. Your photogates could also respond to very short events or even repetitive events for which you need to find a fequency. Of course, there are also commercial programs available for determining frequency of sound, such as one from Vernier for the Apple, or from Covox for the Commodore 64 or Apple. The latter program also allows you to see a diagram of sound shapes on the screen or see the frequency represented as musical notes on a staff.

Frequency programs such as this could be used for telling precisely the relationship between pitch and frequency, testing the frequency discrimination of the human ear in a group of subjects, determining the speed of sound by using resonance tubes, or a quantitative use of the Doppler effect to tell the speed of an object.

Industry, science, and individuals are constantly thinking up new ways to let scientists do the creative parts of an experiment and have computers do the tasks that are dangerous, difficult to do accurately, or too boring.

7

WORD PROCESSING

One of the fundamental tenets of the scientific method is that experiments must be repeatable. If an experiment can't be repeated, the results haven't been validated. Not only does the original researcher have to be able to get the same results again, but *any other scientist* in the same field should also be able to get those same results. To do this, these researchers have to know exactly what you did and know exactly how you did it.

The results of *your* research cannot be effectively communicated to your peers, to science-fair judging committees, or to the world at large unless it is in clear, acceptable form. I've written earlier in this book about how graphs can increase communication of numerical data. However, there still comes the point when you have to write.

Scribbled notes on bundles of paper towels do not make a good lab notebook. A bound lab notebook is a necessity. Handwritten explanations on lined notebook paper do not give your ideas the chance of being accepted that a neatly typed report does.

I cannot type perfectly. I make mistakes. However, I use a word processing program on my computer so that nobody has to know how many mistakes I made or how long it took me to get my writing the way *I* want it to look.

A word processing program allows you to type in your report, correct typographical errors and misspellings, move words and paragraphs around to where they really go *best* (not just where you thought of them the first time), and have the computer printer make a neat final copy with no errors. If you want to change things later, you can load your report from a disk, change it, and let the printer do all the typing again.

There are many word processing programs on the market, varying greatly in price, capabilities, and ease of use. Especially worthwhile are programs that have a great number of features, especially if all of them are obvious after the first ten minutes of use. That's important. You may have the most capable word processing program in the world, but unless it's easy to use, you'll never start using it.

Most word processing programs give you the ability to EDIT, MOVE text from one part of a document to another, COPY some words to another place, SEARCH your document quickly, and FIND AND REPLACE.

That last feature is deceptive. You might think its only use is correcting a spelling you made wrongly throughout your paper. The way I use it is to avoid mistakes. For example, if I have to use the phrase "light-activated chlorophyll electrophoresis" a number of times in my paper, I'll just use an asterisk (*) wherever it's supposed to go. Then I use the FIND AND REPLACE function to replace all asterisks with that troublesome phrase.

Most programs also allow you to format your document for printing: select margins, page breaks, how you want the lines aligned, print style, and headings and footings.

Some programs allow you to create form letters.

Screen from Sierra On-Line's
HomeWord *word processing program.*

This is a feature where you can send the same letter to many recipients, with some things changed for each one, even if it's only the name and address at the top. For a science project, you might find yourself writing numerous letters to large groups of people. Perhaps you might wish to get similar information from some manufacturers or information for a statistical survey in a psychological experiment.

Another feature you may find valuable is a spelling checker, a program that goes through your paper, underlining words it thinks are misspelled. It won't find "hear" when you meant "here," but it will usually pick up typographical errors. The program contains a dictionary to check your spellings. Spelling checkers often come on a separate disk from your word processing program. Most programs include a lot of other features too numerous to go into here. The main thing is to find a program that is easy to use and does what you want it to do.

An interesting newer development for microcomputers is integrated software, in which several modules can communicate with each other. One such program, *AppleWorks*, combines a spreadsheet, a data base, and a word processor. Tables of data from the data base and calculations from the spreadsheet can be moved into the word processor (for final printout in your paper). Although such software takes time to learn, it can be well worth your while if your paper can take advantage of these extra features.

A word processing program will not write your paper for you. What it will do is to make the writing of your paper much easier, neater, and have it come out the way you want it to at the end. Science should enlighten, not confuse.

APPENDIX 1: MORE STATISTICS

Chapter 3 gave you a good start on using accepted measures of how valid your conclusions can be. What I have given there, although fairly straightforward, may be somewhat too strict for your data to satisfy the tests for being "statistically significant." There is a way to claim that validity even if your results are not outside of one standard deviation away from the average. All it takes is a little further treatment of your results—one which *can* be done with a computer but may be just as handy on a calculator. The trouble is, you have to understand what's really going on first.

I have put this in an appendix since it takes a little more algebra and is not as easy as the other methods. You can get by in many cases without these methods, but they may be needed. (A correlation program is available on the disk which can be obtained from the author. See the Introduction.)

THE Z-TEST

This method, called the *z-test*, depends on finding things at the ends of the bell-shaped curve. These ends are called the tails. Sometimes the curves are steeper than usual and the tails really would contain very few items, even in a normal distribution. First, however, we have to decide whether we want to do a *one-tailed* or *two-tailed* test.

HOW MANY TAILS?

It all depends on how you state your hypothesis. If you're just trying to find out if there is any *difference* in your special group, compared to the control, you are really looking for anything at *either end* of the curve—either tail. If you state in your hypothesis that you are looking for only *increases* in something, for example, then you are just looking at one end of the curve—one tail.

Let's use the example where you have fifty pea plants that you have been growing under controlled conditions. For your control group, the average size after two weeks was 10 centimeters (a little under 4 inches) with a standard deviation of ±1.2 cm.

For another group of twenty plants, you used soil to which you had added 10% wood ashes. Their average height turned out to be 11.1 cm. Is this a significant difference? Using our earlier idea from Chapter 4, since 10 cm + 1.2 cm = 11.2 cm, this would seem not to be significant. Let us proceed with the z-test.

The reasons for the following method are too long to explain here. Just take it as a "cookbook method" to use.

1. Find the difference between the average of your control group and the special group. For our example:

$$\text{average}_{\text{control}} - \text{average}_{\text{ashes}} = \text{difference}$$

$$11.1 \text{ cm} - 10.0 \text{ cm} = 1.1 \text{ cm}$$

2. Take the square root of the number of plants in your special group that had the ashes added:

$$\text{plants} = 20, \text{ square root} = 4.47$$

3. Now divide the standard deviation of your control group by the square root of the number of your special group:

$$\text{std. dev.}_{\text{control}}/\text{sqr. root}_{\text{ashes}}$$

$$1.2/4.47 = 0.268$$

4. Now take the difference of the averages, from step 1, and divide that by the result you got in step 3. This gives you a z-value:

$$\text{difference}/\text{result}_3 = \text{z-value}$$

$$1.1/0.268 = 4.10$$

5. Check result with critical z-values in Table 4.

Table 4 Significance Test (Z-Test)

	Critical z-value	
Significance level	One-tailed test	Two-tailed test
0.05 (95%)	1.65	1.96
0.01 (99%)	2.33	2.58
0.001 (99.9%)	3.08	3.30

Our z-value came out to be 4.10, which is greater than the critical z-value for a significance of 0.001. That means, according to our data, and the z-test, the right-hand tail of our curve at 11.1 would actually contain *less than 0.001 (0.1%)* of our expected values. Therefore, you can say that your ash treatment on the soil made a significant difference, with a 99.9% confidence level. If our result in calculation 4 had come out to be, for example, 1.8, you could say the results were statistically significant, but only to the 95% confidence level.

Let's see what would have happened if the standard deviation of our control group had been bigger, say ±3.0 cm. The only things we'd do differently would be steps 3 and 4.

Step 3: 3.0 cm/4.47 = 0.67
Step 4: 1.1 cm/0.67 = 1.64

Now, the z-value is too small to fit even our 95% confidence level. With a standard deviation of ±3.0 cm, an average difference of only 1.1 cm between our control group and the group that we did something special fits well within the range of what we could expect for normal growth patterns for our peas.

Notice that it's harder to meet these level requirements for a two-tailed test. Therefore, it's best to try to state your hypothesis in one-tailed terms. Using our example, you would hypothesize that "Adding 10% wood ashes to the soil will result in *taller* growth for pea plants," instead of saying the two-tailed version: "Adding 10% wood ashes to the soil will result in *different* growth for pea plants."

As you can see, using "statistics," you can make the same figures either be "statistically significant" or not, depending which test you use.

CORRELATION ANALYSIS

In some experiments, you may wish to prove that two sets of data are not different, but instead *go together*, that is, they *correlate*.

For example, you might wish to prove that the Dow Jones stock averages are related to the number of people employed. You would collect your data from the newspapers, taking an average for the month, and then get the employment figures from the U.S. Department of Labor Statistics. They may *seem* to go along with each other, but there is a way to say that the correlation is statistically significant. It involves a bit of math, and the arithmetic is tedious. Naturally, a computer program can do just such a task quickly.

In such a program, you might work with two different types of data. There might be the numbers like the stock market and employment figures above, or there might be "rank ordering." This is the type of data where you say which thing is first, which is second, and so forth.

For example, you might wish to take a list of the top ten students in the school and ask the rest of the student body to rank them as to who is first down to tenth. Then you could ask the teachers to do the same thing. You would run the program and see if student estimations of their peers correlate with teachers' rankings.

After putting in your data, you would be told that there is a POSITIVE correlation, meaning the two sets of data *are* related; or a NEGATIVE correlation, which says the two sets are definitely not related; or that there is no significant correlation, which means that some of the points seem related, others not, and you really can't say.

One word of caution: saying that two things correlate says NOTHING about cause and effect. On some statistical studies, the number of salmon swimming past the fish ladders on the Columbia River has correlated positively with human birth rates in the eastern United States. It really doesn't make much sense to try to prove that one causes the other, but correlations *do* give you a place to start in looking for cause-and-effect relationships.

APPENDIX 2: SOFTWARE COMPANIES

Apple Computer Co., 20525 Mariani Ave., Cupertino, CA 95014

Atari Corp., P.O. Box 61657, Sunnyvale, CA 94086

Broderbund Software, 1938 Fourth St., San Rafael, CA 94901

Conduit, the University of Iowa, Oakdale Campus, Iowa City, IA 52242

Covox, 675 Conger, Eugene, OR 97402

Cross Educational Software, P.O. Box 1536, Ruston, LA 71270

HRM Software, 175 Tompkins Ave., Pleasantville, NY 10570

Ken Hartman, Ames High School, 20th and Ridgewood, Ames, IA 50010

The Learning Company, 4370 Alpine Rd., Portola Valley, CA 94025

Lightspeed Software, 2124 Kittredge St., Suite 185, Berkeley, CA 94704

NCSSM, P.O. Box 2418, Durham, NC 27705

Prentice-Hall, Inc., Englewood Cliffs, NJ 07632

Sensible Software, Inc., 24011 Seneca, Oak Park, MI 48237

Sierra On-Line, Inc., Coarsegold, CA 93614

Spectral Graphics Software, 6333 Pacific Ave., Suite 256, Stockton, CA 95207

StyleWare, Inc., 6427 Hillcroft, Suite 201, Houston, TX 77081

Vernier Software, 2920 SW 89th St., Portland, OR 97225

Wadsworth Electronic Publishing, 555 Abrego St., Monterey, CA 93940

GLOSSARY

Algorithm - The method by which a program figures out a task. The *algorithm* for finding the area of a rectangle is to multiply the length by the width.

Average - In statistics, finding the mean value by adding up all items and dividing by the number of items.

Data Base - An electronic filing system. A program which allows words and numbers to be sorted many ways and selectively printed out.

Indexed Array Sorting - A way of organizing a list of items in a computer array by not moving the individual items but keeping another array which is a list of the order things are supposed to be in.

Program - A sequence of instructions to a computer that tells the computer what operations to do in what order. The program may also contain the data on which the operations are to be performed.

Significant Figures - In science, the number of digits in a number that have been measured accurately. "It's exactly 342 miles" uses a number with *three significant figures*. "We're about 40 miles from New

York," uses a number which has only *one significant figure*.

Simulation - In computers, a program which gives the same results as a real event. In some cases, the results are expressed in numbers; in many cases, graphics give a pictorial representation of the activity.

Smart Tools - Tools (such as a thermometer) connected to a microprocessor so that not only is the measurement taken but the results are reported digitally and recorded by the computer.

Sorting - Organizing information into some kind of order.

Spreadsheet - An arrangement of information, usually numbers, in columns and rows that facilitates manipulation of the numbers.

Standard Deviation - In statistics, an average difference from the average that will contain 68% of all the figures of a random sample.

Statistics - A mathematical treatment of numbers so that predictions can be made of how the group will act in the future.

Thermistor - A resistor which increases resistance as it gets hotter.

Variance - In statistics, the square of the standard deviation.

SELECTED READING

Ahl, David H., ed. *Computers in Science and Social Studies*. New York: Creative Computing Press, 1983. Many programs from early microcomputer days. Most of them use no graphics and so are "generic" for a lot of models.

Beller, Joel. *So You Want to Do a Science Project!* New York: Arco, 1982.

Brain Bank (Brain, David A., and others). *The BASIC Conversions Handbook for Apple, TRS-80, and PET Users*. Rochelle Park, N.J.: Hayden, 1981. A required book for converting programs among these machines.

Carmony, Lowell A.; McGlinn, Robert J.; Millman, Ann Miller; and Becker, Jerry P. *Apple Pascal: A Self-Study Guide*. Rockville, Md.: Computer Science Press, 1985.

Clark, James F., and La Barre, James E. *Advanced Structured BASIC*. Cincinnati: South-Western Publishing, 1985.

Crider, Bill, ed. *BASIC Program Conversions: How to Convert Programs from One Computer to Another.* Tucson, Ariz.: HP Books, 1984. Contains conversion principles and language changes for IBM PC, PCjr, Apple II+ and IIe, Commodore 64 TRS-80 models III and 4 and the Color Computer.

De Jong, Marvin L. *Apple II Applications.* Indianapolis: Howard W. Sams, 1983. Very useful for interfacing with computers and machines or sensors.

Dowdy, Shirley, and Wearden, Stanley. *Statistics for Research.* New York: John Wiley, 1983. A bit heavy at times but does stress specific statistics applications for science.

Goodwin, Peter. *Physics with Computers.* New York: Arco, 1985.

Gray, Susan H. *No-Frills Statistics: A Guide for the First-Year Student.* Totowa, N.J.: Rowman & Allanheld, 1984. Very useful for explaining the techniques used in Dowdy and Wearden.

Kosniowski, Czes. *Fun Mathematics on Your Microcomputer.* New York: Cambridge University Press, 1983. A neat book to play with. The math projects will give you a lot of ideas for using math in science.

Lampton, Christopher. *BASIC for Beginners.* New York: Franklin Watts, 1984.

Lampton, Christopher. *Graphics and Animation on the Commodore 64.* New York: Franklin Watts, 1985. Creating graphics on the Commodore is very difficult to do on your own. This book helps make sense of the PEEKs and POKEs.

Lampton, Christopher. *Pascal for Beginners.* New York: Franklin Watts, 1984.

Luehrmann, Arthur, and Peckham, Herbert. *Apple Pascal, a Hands-On Approach.* New York: McGraw-Hill, 1982. If you're going to get into Pascal, you'll need several books. Luehrmann's, Carmony's, and Lampton's are the most readable and useful I've found.

Schulman, Elayne; Craigo, Ken; Griffiths, William F.; and Megna, Denise. *Science Projects with Computers.* New York: Arco, 1985. Especially good for easy projects and computers without graphics.

Webster, David. *How to Do a Science Project.* New York: Franklin Watts, 1974. Good general work on project design, process, and presentation.

Wolfe, Philip M., and Koelling, C. Patrick. *BASIC Engineering, Science, and Business Programs for the Apple II and IIe.* New York: Brady, 1984. The program listings are helpful, but you are supposed to know what they're to be used for.

INDEX